The Territory of Metaponto 1981–1982

The University of Texas at Austin

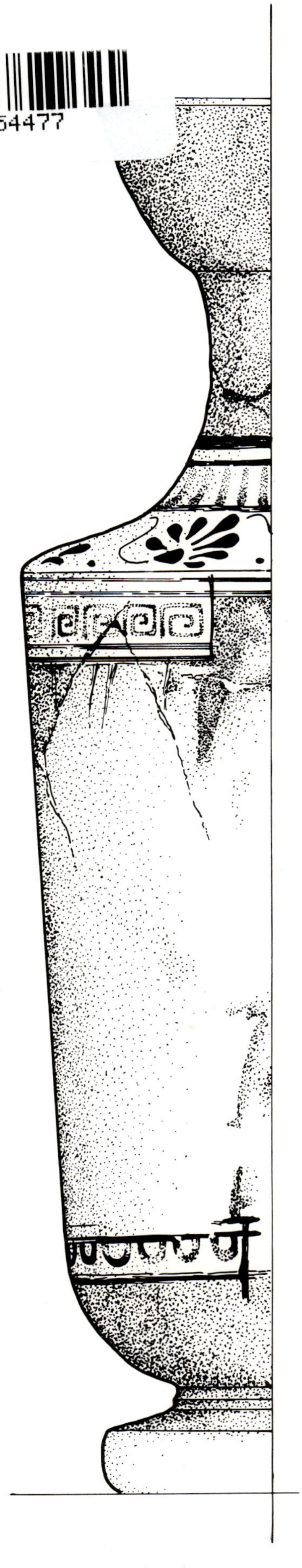

With Gratitude

This year the Metaponto project enters its tenth year. It has grown, as the years have passed, both in complexity and scope. What began in 1974 as the excavation of a single rural site has become by stages a multidisciplinary study, of a territory consisting of hundreds of sites—one that spans the whole of its long and sometimes dramatic history, with broad implications for the social and economic development of the Mediterranean area. This steady growth has depended entirely on two conditions: sustained and generous support from donors and foundations, and the ability and enthusiasm of the participants in the project.

It is no exaggeration to say that the 1981 and 1982 campaigns were the most varied and productive yet. They were made financially possible by grants from the National Endowment for the Humanities (which renewed its support in May, 1982 for an additional three year period), the James R. Dougherty Jr. Foundation of Beeville, Texas, and the Brown Foundation of Houston. The University of Texas, through the Institute of Classical Archaeology, has provided a basis for this organized research project. I am extremely grateful to the private donors, whose support, both financial and moral, has been unwavering over the years. They are individually acknowledged on the inside cover, together with the students and professionals who have shared their energies and varied skills for the past two years. In addition to the donors listed there, I would like to acknowledge the Swedish School in Rome, and the Constantin Fund of the University of St. Thomas, Houston, for aid in support of student and faculty participation in the project during 1982.

This project was first conceived in the generous atmosphere of international collaboration, which has always characterized the Superintendency of Antiquities for Basilicata, directed successively by Professor Dinu Adamesteanu, Dr. Elena Lattanzi, and Dr. Angelo Bottini, and which has continued to make this region a focus of innovative archaeological research. This close working relationship with our Italian colleagues was broadened still further when, five years ago, paleobotanist Dr. Lorenzo Costantini of the Ministry of Cultural and Environmental Resources joined forces with us, adding a whole new dimension to the project.

Our principal base in the territory has always been the experimental farm administered by the Region of Basilicata at Pizzica-Pantanello. We are grateful as ever for the hospitality. I am sure that in the name of international collaboration we have received far more than we could ever hope to repay, or even adequately acknowledge.

Joseph Carter
Director
Institute of Classical Archaeology

Rural Archaeology at Metaponto: Aims and Results in 1981 and 1982. *A Synopsis*

JOSEPH CARTER

^

East-west wall (foreground) on an extensive Greek (and later Roman) farmhouse site, near the spring at Sant'Angelo Grieco. It is located about 12 kilometers to the northwest of the ancient city, below modern Bernalda. The site was noted by the survey team in 1980 and again in 1981. Surface finds included archaic and classical Greek ceramics, and pottery of the first century B.C. and first century A.D. Excavation revealed that foundations of a building of the fourth century B.C. (including the wall in this view) were reused in a large building of the late Republic and early Empire. Century old olive trees protected the central section of the complex from destructive plowing.

Our major objective, almost from the start, has been to trace the vicissitudes of the rural population, the inhabitants of the territory surrounding the important Greek colonial city of Metaponto, throughout the period of Greek and Roman occupation of the area. These ten or eleven centuries (roughly 700 B.C. to 400 A.D.) encompass major transformations in which our subjects must have been actively involved. Their role in this history, here as elsewhere, has never received even remotely the same attention as that devoted to the urban populace. It was, as we hope to show, an important one.

The investigation begins, chronologically, with the arrival of merchants and adventurers from Greece who, our scanty historical sources inform us, seized land from the natives, but not without a fight. What immediate changes did the arrival of a highly organized group of farmers produce on the landscape? This initial phase was followed by one of rapid growth and consolidation in the archaic and classical periods, which is well documented at Metaponto, above all by the Italian excavations of the past twenty years in the urban center. A second phenomenal period of growth occurred in the late Classical and early Hellenistic periods (350–275 B.C.). If the advent of the Greeks occasioned a major shift in settlement patterns and land use in the territory [see our 1980 report], a second coincided with that of the Romans and their domination of the area. Once again the historical sources give only the slightest hint of what must have been profound changes as agriculture gave way to pastoralism. Was the Roman influence felt immediately? How rapidly was the land impoverished? and how directly were human agents responsible for the decline mentioned by the ancient writers? These questions have required close collaboration between specialists, in such areas as faunal and pollen analysis,

Continued next page

PHOTO OPPOSITE PAGE >

Excavation in the area of the Greek spring-sanctuary, 1982. The "well-point" pumping system, which made excavation below the water table possible, began to lose the battle with the water just as the excavation reached its most critical phase. Here, the specialized excavation team works frantically to make it operative again. The solution adopted, that of forcing water under high pressure through the "points" to clean them, worked and the excavation was completed on schedule.

and classical archaeologists, which I believe has been illuminating as well as solidly productive for all involved.

Something of the importance and the variety of rural settlements has been learned through excavation of selected sites by the Superintendency of Basilicata and its collaborators, including the University of Texas. In fact, this was the principal activity of our mission during its first six or seven years. Examples of the major types of settlement in the periods mentioned above have now been brought to light—farmhouses (Fattoria Fabrizio, San Biagio, Pizzica, Sant'Angelo Grieco), ceramic industries (Pizzica and Sant'Angelo Vecchio), necropoleis (Saldone, Sant'Angelo Vecchio), and sanctuaries (Pizzica, Incoronata). Much of our time and energy for the past several years has been directed at preparing this material for publication. This aspect of the project, which includes special studies of Greek and Roman ceramics, small finds, and architecture, analysis of organic materials, as well as the detailed description of these sites, has been closely coordinated with excavation.

Excavation has been and hopefully will continue to be an important part of the project, but it yielded some ground in 1981 and 1982 to the survey, and to paleobotanical and faunal studies. The survey team, which was formed in 1981, is making intensive surveys of selected areas of the territory. In autumn 1982, it completed a "blanket" survey of a forty square kilometer area, stretching across the territory from the Basento to the Bradano Rivers. The surface covered is the equivalent of about an eighth of the territory of Metaponto. Every ancient site was fully documented and plotted on a detailed map. The total number of sites, after something less than six month's work by the team of five, stood at four hundred and fifty-three—many times more than the most optimistic prediction. If this density, ten sites per square kilometer, proves to be characteristic of the whole territory, an estimate of 2000 separate family units should be about right for the heyday of rural settlement, in the late fourth century B.C.

This new information is of great interest in itself. With a population of, say, eight to ten thousand living outside the city walls, there can be no doubt that "Metaponto" meant the countryside, as well as the urban center. Even more interesting are the further questions posed by the survey results. The ceramic evidence indicates that occupation of the territory in periods immediately preceding and following the late fourth century B.C. was relatively light. The theories that are often adduced by archaeologists to explain a sudden rapid increase in population—the "agricultural revolution"—or the sudden decline—"invasion", "civil strife"—are not really satisfactory. So little is known of ancient agriculture, especially pre-Roman, that "revolution" in this context is a meaningless term.

Ancient writers dealing with the period, however, have provided us with some potential as well as actual invaders. Cleomenes, for example, was summoned from Sparta by Taras (modern Taranto) to lead the resistence of the Italiote Greek cities, including Metaponto, against the Roman threat. When Metaponto, according to the historian Diodorus Siculus, demurred, Cleomenes forced the barbarian Lucanians to invade the *chora* (territory) of Metaponto, demanding six hundred talents of silver and two hundred girls from the best families as the price for "saving" them—an indication, by the way, of the wealth of the city at that time. Cleomenes' activity must be placed shortly before 300 B.C., but the ceramic evidence from the territory (which requires further extensive study) seems to indicate that the population level was still high in the early third century B.C. So, Metaponto must have recovered.

The beginning of the decline seems rather to have coincided with the last of the "condottieri", the Epiriote king, Pyrrhus. His expedition to Italy (282–275 B.C.) and attempt to halt the Roman expansion into the south were failures. No source, however, mentions him in connection with Metaponto. It is satisfy-

ing to be able to find corroboration and support for the archaeological data in our own meager records, but shouldn't we, after all, also be asking more fundamental questions—such as, what weakened Metaponto and similar cities to the point where a Cleomenes or Pyrrhus could have his way with them? Should we be looking for a single event or a unique cause for what must have been a complicated process?

In recent years some archaeologists of the classical world have been turning increasingly to the physical sciences (paleobotany, palynology, paleozoology and geology) for new evidence, and to the social sciences (anthropology, geography, economics) for interpretive models to broaden their understanding of the complex historical situations they confront. We are moving in this direction in our study of the territory of Metaponto, as this and earlier reports illustrate. The results of the survey represent a major advance, since they provide us with the first requisite of statistical work: a substantial "data base" containing the sites and their detailed descriptions. One goal that now seems within reach is an accurate estimate of the size of individual properties in the territory at different periods—a key to understanding the economy.

Our other major advance during the past two years relates directly to the principal activity of the territory—farming. Excavation at Pizzica in 1981 was closely coordinated with a faunal study. It produced impressive evidence of a major change in the animal populations and land use for this area under the Romans of the late Republic (150–50 B.C.). In 1982, excavation had as its principal goal the recovery and interpretation of the unique deposits of organic material preserved in the late fourth-early third century B.C. collecting basin at Pizzica. This was a formidable challenge as the site was largely under water. These conditions were responsible for the extraordinary preservation of the material, but made it impossible to conduct the precise and clean excavation required to produce full and coherent results. It took some considerable technical innovation, coordination of specialists (in paleobotany, excavation, hydraulic engineering, and site documentation), and a great deal of back-breaking work in a breezeless, shadeless pit where the temperatures habitually reached 105°F. Under such conditions not only do team members need expertise and imagination, they have to mesh on a human level with their coworkers. This campaign was a thorough success, and, I might add, one of the great satisfactions of my decade at Metaponto.

The results are an immense amount of information not only about the kind of agricultural plants grown here between 350 and 275 B.C., but also about the environment as a whole. As studies proceed, it will be possible to describe the territory in greater detail than has ever been previously possible, and it places in our hands another crucial set of data. With these, the remarkable, contemporary, documentary evidence (available in the Heraklea Tablets) and the results of the pollen analysis, we will be well along the way to defining the relation between man and the ecosystem, which may well have been a decisive factor at that turning point in the history of Metaponto.

Any disciplined investigation of a complex problem is a time-consuming business, with many unavoidable and tedious details, with sacrifices and satisfactions, moments of near triumph and despair, but one element sustains a project over the long haul and that is the excitement of the chase, the hunt for clues to a solution. The Metaponto project is no exception. Ten years of work have brought us a long way toward the realization of our initial goal—a fuller understanding of a little known aspect of the ancient world. We might have expected the excitement to have tapered off some. On the contrary, in recent years it has been steadily increasing, as both the number and variety of clues increase, and we realize that the initial problem—largely one of description and documentation—has been subsumed by the larger search for the reasons why a civilization rose and declined.

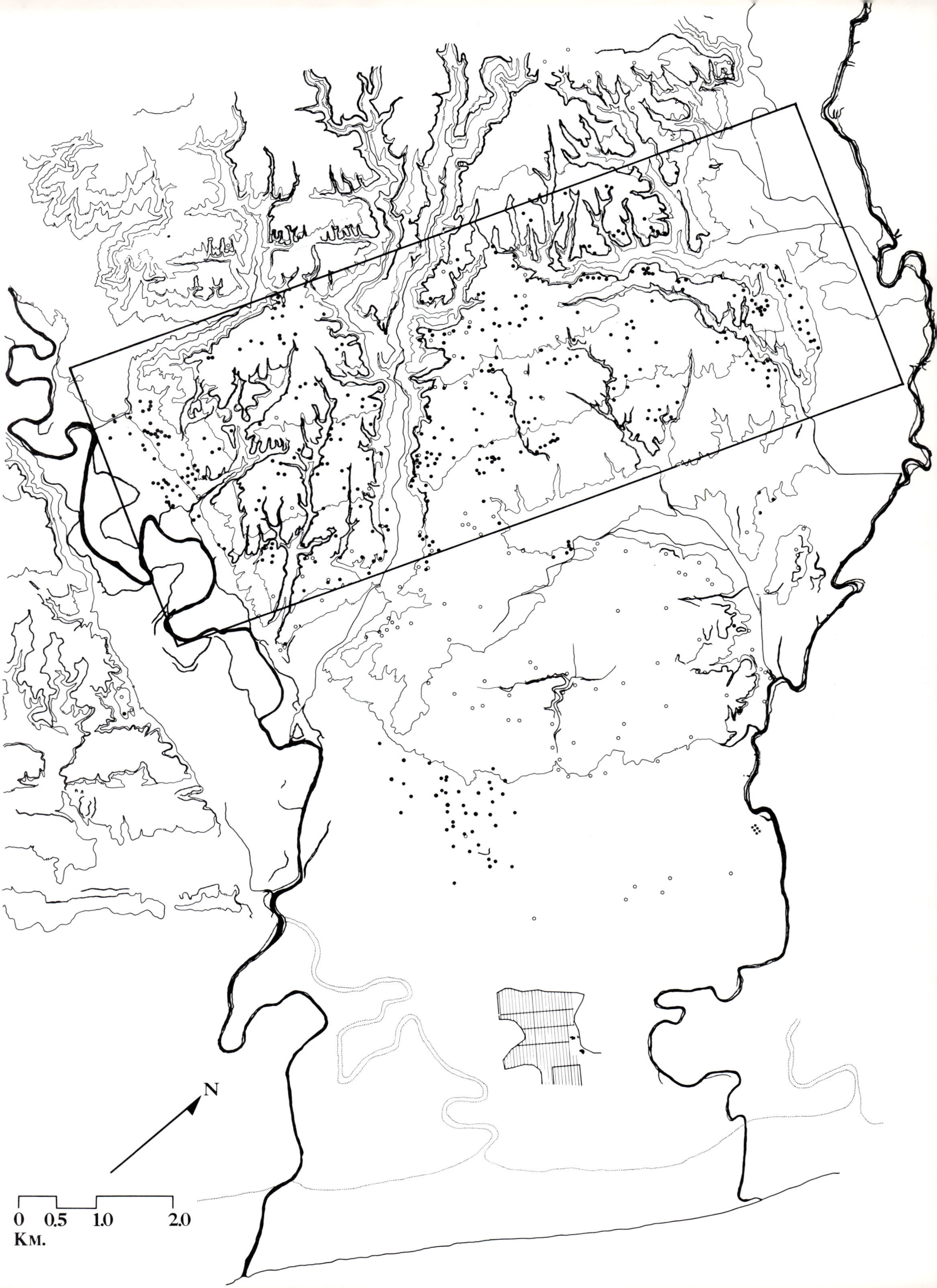

N
0 0.5 1.0 2.0
Km.

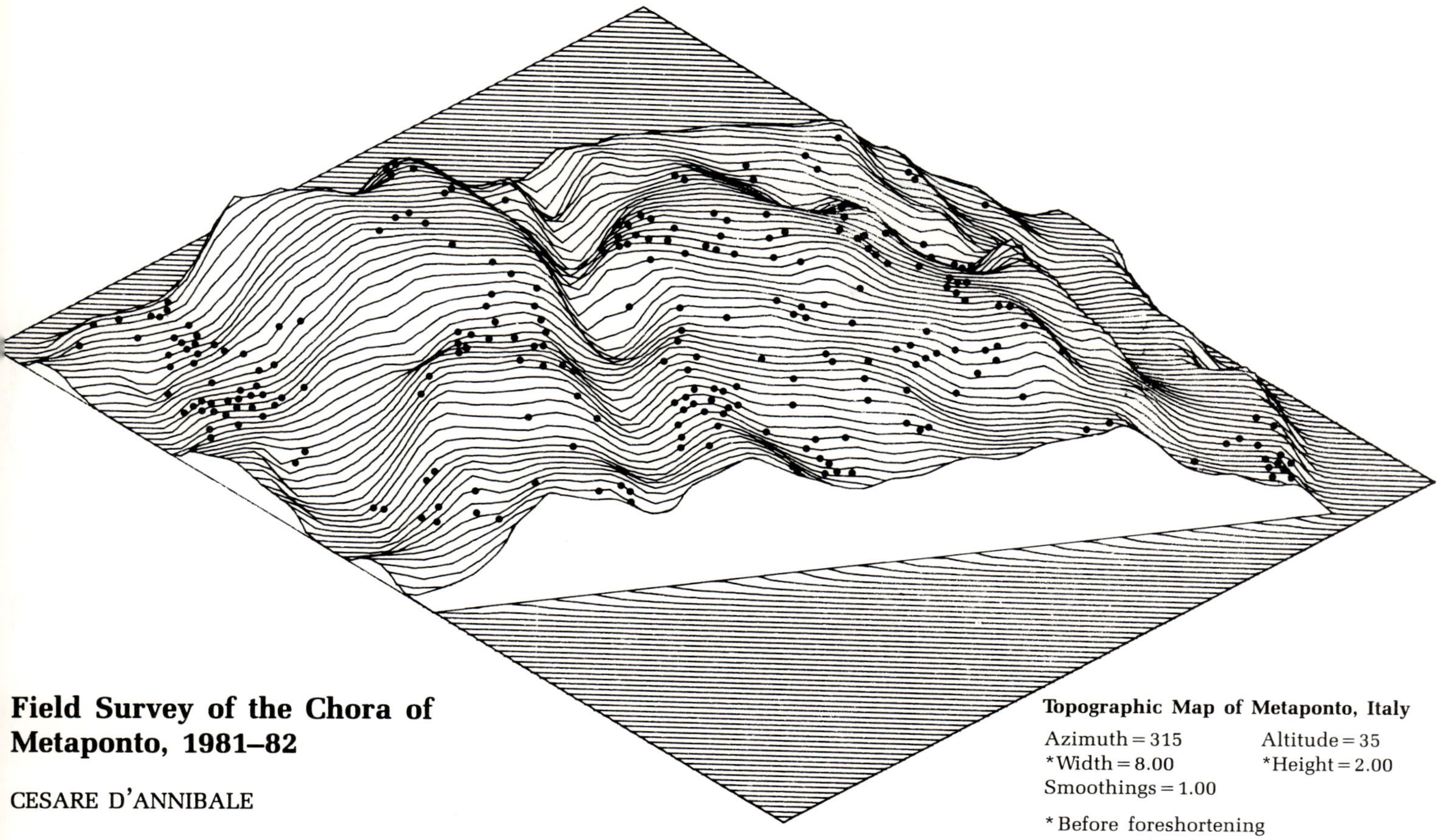

Topographic Map of Metaponto, Italy

Azimuth = 315 Altitude = 35
*Width = 8.00 *Height = 2.00
Smoothings = 1.00

* Before foreshortening

by Jim Abbot and Sandra Langston

Field Survey of the Chora of Metaponto, 1981–82

CESARE D'ANNIBALE

Aims

An intensive survey of the Territory of Metaponto between the Basento and Bradano rivers was initiated in June, 1981. Its primary aim is to discover the actual distribution of ancient sites and complete a census of the territory by period from the earliest settlement (in the neolithic period) down to modern times, and to assess the changes in settlement patterns that occurred in major occupation phases. The survey includes a number of related research projects.

<

Plan of the territory of Metaponto between the Bradano and Basento Rivers. Below: the ancient coastline and river beds, indicated with dotted lines, and the plan of the ancient city, as it has been revealed by the large scale excavations by the Superintendency of Basilica, (1964 to present). Middle: cluster of black dots indicates ancient sites discovered in the Pizzica survey, all in a radius of one kilometer from the Pizzica sanctuary. Top: the rectangle encloses the area of the forty square kilometer transect, which was the survey team's principal object in 1981 and 1982.

There is, for example, a geomorphological component, which so far has focused on the relationship between the much discussed "division lines", the primary geological features, and the settlement patterns in the territory. The data from the survey is being analyzed for the light it can shed on specific historical problems. For one, a study is underway concerning the distribution of sites in the fourth century B.C. (the period of maximum density), in relation to the Heraklea Tablets in an attempt to determine the size of lots associated with individual farmhouses. A project for the near future is to move the survey inland to the contact zone between Greek and indigenous populations on the fringe of the territory to investigate a relationship of fundamental importance in colonial history. The unusually high density of sites in the territory is making it possible for the survey team to construct a very rich "data base", with wide applications in a variety of disciplines.

Continued next page

Λ

A computer graphics representation of the forty square kilometer transect of the territory, between the Bradano and Basento Rivers, (six kilometers inland from the ancient city) covered by the survey teams. The relief has been enhanced, and the plan rotated, so that the line of sight is 45 degrees to the west of true north. This is the approximate orientation of the division lines of the territory. The black dots indicate ancient sites. About eighty percent of these were habitations. The rest were tombs (which are normally found in close proximity to farmhouses). In this computer drawing, the concentration of sites on the high terraces, along the rims of valleys, and in the river bottom land, is evident.

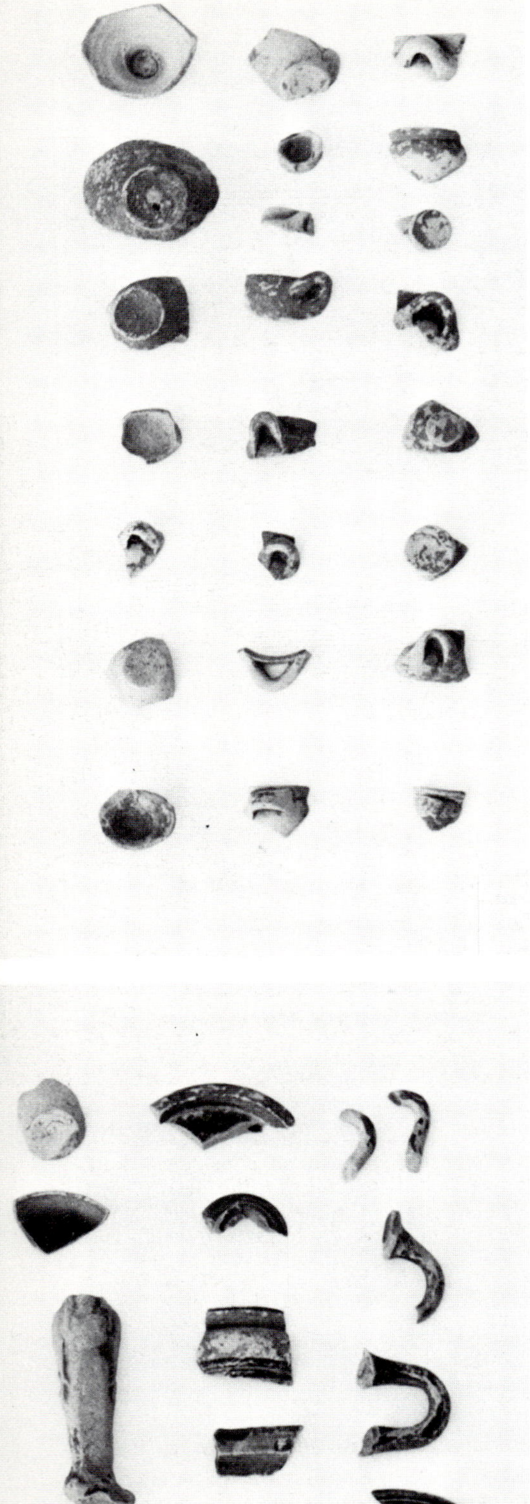

The Research Unit

The survey, it is hoped, will be a long term project, so that total blanket coverage of the Metapontine territory can be made. This is in the realm of possibility. It should extend inland at least 17 km to the point where the level coastal plateau and the hilly terrain of the interior meet. The average width of the territory is 10 km, (but it broadens as we move inland). This geographical unit corresponds to the limits of the ancient territory and was culturally, politically, and economically indivisible from the urban center. (A similar unit lay between the Basento and Cavone rivers, so that the research unit is strictly only the older half of the territory of Metaponto.) A cross-section, or transect, 4 km wide and 10 km long, was designated in 1981 as a first priority. It runs between the two rivers, roughly parallel to the coastline, and is about 8 km from the ancient city at the nearest point. The terrain in this cross-section is representative of the rest of the territory. It can be divided into three distinct topographical areas: the river plains and bottom land along the tributaries, the sloping sides of valleys, and the high ground of the marine terraces.

Procedures

The field crew consists generally of four or five members walking on the average 10 meters apart. Spacing of the field crew combined with good surface visibility makes it relatively easy to identify the smallest site, usually a tomb site, 5 to 50 meters in diameter. At most sites the surface indicators consist of fragmentary roof tiles and sherd scatters, the average farmsite having a diameter of 30 meters. Sites are recorded by stating their dimensions, density, and other noticeable features. Both black and white and color photographs are taken for each. Most of the diagnostic material and a tile sample are collected. All pottery is catalogued according to ware types.

Progress, 1981–82

In 1981, 139 sites were recorded, the majority of them for the first time. [Surveys carried out from 1966 to 1973 by the Superintendency and its collaborators focused on other areas.] The greatly increased surface scatter from ancient sites in certain areas (such as Cogno del Prete, where no sites had been noted in the earliest surveys) reflects recent, intensified agriculture. During 1982 the survey season filled the months of June and July, and was extended for the first time into the fall. The total number of sites reached 453.

In 1982 a total of 251 sites were found within the 10x4 km transect across the territory. No large portion of land in this zone was unoccupied, especially in the Archaic and Classical Greek periods. The density of ancient sites is far greater than any earlier estimate. Site location seems to have been determined primarily by accessibility to water. This is most evident around springs at San Angelo Vecchio, San Angelo Grieco, and the southern end of the Venella (around San Biagio). At these locations we see a dense clustering of sites, including large farmsites, their associated buildings and burial grounds, kiln sites and sanctuaries. In other areas, such as around the Fosso dell'Acqua Fetente, Fosso San Marco, and other small streams or gulleys that bisect the plateau, sites can be found on the edges overlooking the gulleys. Judging by the extraordinary amount of votive cup fragments found on the surface, there is a high probability that Site 270 in the Giampasquale locality, Site 334 on the lower flanking terraces of the Pantano Venella and Site 397 near Masseria Avinella are small sanctuary sites. The remains of what appear to have been six

<

An unusually interesting collection of material, from site 397 (located along the valley of the Venella, not far from San Biagio). It consists almost entirely of fragments of small votive cups, and of fine archaic pottery, and includes a (headless) statuette in the Daedalic style. This site in all likelihood was a small rural sanctuary, of about 600 B.C. There was no evidence of later occupation.

^

An imposing fragment of a medieval wall, Pietra San Giovanni, in the territory of Metaponto. This landmark, which has been an object of study by the Superintendency, was included in the "blanket" survey by the University of Texas team, as site 300. Nearby is evidence of a bronze age habitation. These sites represent the chronological extremes of the survey; they span roughly 3000 years of history.

kilns were also identified, primarily by the slag and some misfired sherds on the surface, near Masseria San Angelo Vecchio, Spineto, and on the terraces of the Bradano river below Masseria San Marco.

The number of sites which contained lithic remains was rather small: only 18. At only one site, 271, are the remains dominant. These and others have not yet been dated. A surprising quantity of Bronze Age/Early Iron Age impasto sherds were found in 1982. A total of 7 sites contained this type of material, in just three areas of the territory: Giampasquale, overlooking the Fosso San Marco, below Masseria San Marco on the Bradano river terraces, and in the Avinella area, overlooking the Pantano Venella.

The Greek phase of occupation was extremely well documented. Sherds from this period were present on at least 227 of the 251 sites. Greek sites blanket the entire survey area, indicating the intensive activity in the countryside immediately outside Metaponto.

Human activity in the Roman period is also well documented—49 sites clearly have Roman remains. Most of this material is found in association with Greek. The most impressive Roman site, 230, is located below Masseria San Marco on a Bradano river terrace.

Evidence for Medieval occupation of the territory is found at only 7 sites. Though this number is small, the quantity of material found, especially from two sites, 140 and 300, indicates that these at least were quite important during this period. Site 140 is located near Masseria San Marco and occupies a long spur bounded on one side by the Fosso San Marco, and on the other by the Bradano river. Site 300, known as Pietra San Giovanni, also has the same type of topographical setting, but is located on a more imposing spur, jutting out into the valley of the Venella. This site is known primarily for the section of wall that still stands to a height of 4 meters, and which may have been part of a defensive wall whose foundations can still be traced on the extreme tip of the plateau spur.

Pizzica Survey

A limited objective of the survey, initiated and completed in the fall of 1982, was the investigation of an area surrounding the Pizzica sanctuary in order to establish a clearer context for that major excavation site. An area with a radius of one kilometer, extending from the sanctuary, was chosen. This area, situated on prime agricultural land, has been subjected recently to extensive agricultural activity which has and is changing the original appearance of the terrain through the use of the 1.5 meter plow and the bulldozer. Not surprisingly, most of the sites found were heavily disturbed, if not totally destroyed.

Within the area surveyed, fifty-three sites were found—a very high concentration. Almost half of these can be identified as single tombs or groups of tombs, marked primarily by the presence of limestone slab fragments from sarcophagi, some still preserved in their original size, and by pan tiles, another common feature associated with local burials. The necropolis to the northwest of Pizzica, jointly excavated by the University of Texas and the Superintendency, is just one of the many burial grounds surrounding the sanctuary. Evidently the most extensive burial areas were located immediately to the north and east of Pizzica. [Here, in the 1970's, hundreds of tombs were excavated by the Superintendency.]

The remaining sites surrounding the sanctuary consisted of farmsites and small sherd scatters, perhaps belonging to smaller associated farm buildings. A total of 12 farmsteads were identified from surface remains. They were distinguished by heavy concentrations of ceramic material, predominately tile fragments, covering an area averaging around 45x45 m. Also visible and indicative of this type of settlement were the river stones, sandstones, and conglomerates used in the foundations.

Preliminary dating indicates that 28 of the 53 sites found in the Pizzica area belong to the Archaic and Classical Greek periods. The number would presumably be higher if one includes the many tombs which had no pottery, but which consisted solely of pan tiles and limestone fragments of the sort often found in association with Greek pottery. The only other period represented is the Roman. Roman sites numbered only five, three of which were probably farmsteads.

^ Site 429, of the Pizzica survey, came to light after fall plowing. It is located in a field of the experimental farm Azienda Pantanello. In the background is the dormitory of the school of modern agriculture techniques recently opened by the Region of Basilicata.

The distribution of sites by period

The initial period of expansion into the territory, and that of Metaponto's greatest fame, is the Archaic period, particularly the second half of the sixth century B.C. Few strictly Archaic sites besides tombs have ever been found. The majority of sites with large amounts of pottery belong to the late fourth and early third century B.C. This flourishing period however, was short-lived. Ceramic material securely datable to the second half of the third century B.C. and later is relatively scarce. Sites of the Roman period are much less numerous and do not cluster.

It seems that Archaic sites were reused in later periods since a number have a surface scatter of Archaic, Classical, and more rarely, Roman pottery. These sites must have been the most attractive locations for a farmhouse. The survey's results were tested in 1981 by excavating a farmsite with this particular mixture of pottery at Sant'Angelo Grieco [see below]. The excavation results revealed that the Roman farm took advantage of the earlier Classical and Archaic remains as a source of building material and even incorporated foundations of the earlier period into the structure. The chronological sequence of pottery visible on the surface conformed exactly to that of the excavated structures.

Environmental factors and settlement patterns

All three topographical areas were heavily settled in the late Classical period. In the Archaic and Roman periods evidence for settlement in all three areas is also apparent. Availability of water [as discussed in the 1980 report] is the major determining factor in site location. The terraces, dissected by tributary systems emptying into the Bradano and Basento rivers, were and still are dotted with perennial springs. These springs occur at the contact point between the sandy surface soil layer and the underlying clay layer. Sites around the springs appear to cluster and give the impression of having belonged to small agricultural communities. These concentrations seem also to have stimulated industrial activities, as evidenced by the remains of ceramic kilns. On the plateau many sites are located on ancient marine beachlines left behind by the sea as it retreated some 200,000 years ago.

The most serious problem for the student of ancient settlements in the area is, ironically, the acceleration of modern agricultural development as increased irrigation brings more land under cultivation. Extensive use of the 1.5 meter plow and heavy farmmachinery to cut away hillsides and create gentler slopes are erasing sites which were clearly visible on the surface only 10 years ago.

A Note on the Analysis of Metapontine Roof Tiles, 1982

ROBERT FOLK

In the summer of 1982 a systematic study of roof tiles from various sites in the territory of Metaponto was begun. The first step was to develop a descriptive classification whereby tile properties could be described objectively. Tiles have two basic properties, shape and composition. A symbolic method was developed for describing as objectively as possible the various aspects of shape (character of rim, curvature of the tile, and thickness), and composition in terms of color, and visible grit or temper content. Color is dependent on firing temperature and the chemical composition of the clay. For example, inclusions in the grit of calcium-bearing minerals such as calcite imparts a yellow color. The blue clays, "argille azzure," in the vicinity of Pisticci [at the northwestern extreme of the territory], contain abundant fine dolomite (Ca-Mg carbonate) which burns out in the tile to form a hole with a yellowish halo. Other sorts of temper include rounded quartz sand grains, various metamorphic rock fragments, crushed older pottery or ceramic material, and other minerals. Tomb tiles have a very specific composition: they contain abundant, bronze colored flakes of biotite mica, and must have been imported from some volcanic area (perhaps from as far as the area around Vesuvius).

This study is in its infancy. The major objectives are to construct a chronology for tiles analogous to that which exists for the pottery. This would make it possible to date those sites located in the survey which have only a limited amount of associated pottery, and sometimes none at all. As there are a number of types, distribution maps of the territory by period would be that much more informative. Further studies may enable us to identify specific sources for the clay or grit, thereby pinning down the sources of the wares. With this additional data on each site in the survey it may be possible to trace the distribution of tiles from the various centers of production, both within and outside the territory. At present raw data have been collected from a number of habitation sites located by the survey team. Samples from selected ones will be thin-sectioned for study. What is needed now are tiles from sites of known and very limited age range so that the chronology may be firmly established. A series of tiles from stratified levels at Pizzica, soon to be analyzed in the laboratory, should help to provide a solution.

Continued next page

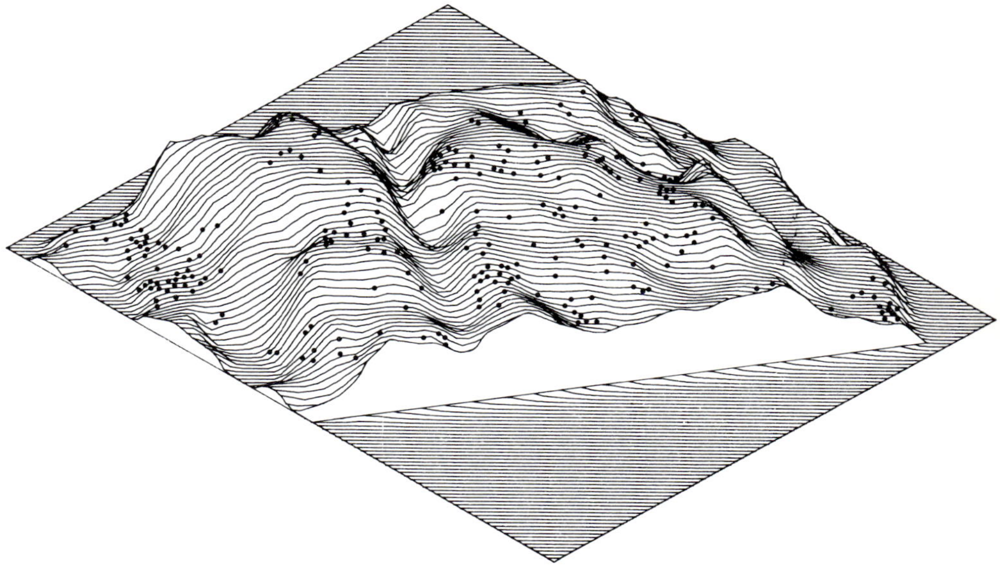

Computer Analysis and the Field Survey

RICHARD JONES *and* MARK REED

The results of the 1981–82 field survey of the territory of Metaponto have raised a significant problem: how to interpret the enormous amount of data produced by survey archaeology. Computers, now gaining wide acceptance among classical archaeologists, are very effective tools for analysing these large quantities of data. The key to analysis though, here as well as in excavation, is to ask the right questions. For example, the literary sources suggest that the Metapontine countryside suffered a change in the late-Classical or early-Hellenistic period from agriculture to grazing. Is there any evidence in the survey results indicating that such a change took place? If single family farms gave way to ranches, this should be evident in the increase of larger sites in the Hellenistic-Roman period (measurements being based on the size of the tile and sherd scatters found on the surface).

To test the hypothesis that site size varied through time, the chi (x^2) square statistical analysis was computed. The sites were arranged in six groupings according to size. The time periods used were Classical, or fifth-fourth century (as evidenced by the presence of black-glaze pottery); the Hellenistic-Roman (grey ware); and the post-Hellenistic (evidenced by the presence of Arretine and/or medieval pottery). The computed chi (x^2) square value of 29.83 with 10 degrees of freedom suggests that there is a statistical difference (.05) between size and time period. In other words, site size varied through time in a significant way. As can be seen from the table, the number of sites in the classical period with a size of 151–625 square meters, 58, dropped substantially to two in the Hellenistic-Roman, whereas the number of larger sites tended to decrease only slightly or remain the same. Altogether the number of sites declined dramatically.

In order to further test the hypothesis, we checked to see if site size varied with distance from a source of water "controlling" for time. This is a method of testing one variable against another to see if their correlation is affected by a third variable. An additional chi (x^2) square was set up, matching groups of sites based on size (the same groupings as in the table) and selected distances from a water source. We tested sites in the Classical, Hellenistic-Roman, and post-Hellenistic periods.

For the Classical period, the computed chi (x^2) square value of 42.67 with 25 degrees of freedom suggests that there is a statistical difference (.05). Unfortunately there is not as yet enough data to say this conclusively for the other two, later periods. It seems that the larger sites tended to be at greater distances from water, a curious result since the larger—and we assume—more prosperous sites ought to have been closer. However, if the economy were based on grazing rather than farming, and a significant number of smaller sites occupied by shepherds rather than farmers, then distance from water might have been a matter of convenience instead of necessity. Another factor affecting site location, which has to be taken in account, is the presence of springs which are no longer visible.

In the future we hope to continue using statistical methods to aid reconstruction of the territory. It is possible that we might establish a statistical correlation between a site's size and its elevation, or its date and its relation to the supposed division lines, both of which might help us determine the type and distribution of farms in the Classical period.

Time-----> Site Size (square meters)	Fifth-Fourth Centuries	Hellenistic-Roman	Post-Hellenistic
1–150	39	11	2
151–625	58	2	4
626–1225	51	10	7
1226–2000	46	10	5
2001–3000	23	10	2
>3000	11	11	4

The Text and the Archaeology of Rural Settlements

JOSEPH CARTER

Classical archaeologists have from the beginning made use of the written record in interpreting the results of excavation. The observations of a Pausanius, Pliny, or a Strabo, however prosaic (and from time to time misinformed) they may be, have in many cases provided the major clues to the identification of a site, the interpretation of an urban center, or the attribution of a work of art. This relationship has served the archaeologist of the Greek and Roman world well, as long as he has confined his interest to areas of mutual concern—the life and culture of the city, the masterpieces, the tourist attractions. What happens though, once he moves out through the suburbs and into the classical landscape? There are few guides. The Roman writers on agriculture are an important but very limited resource, both in the time period they cover and the range of their interests.

In the particular case of the territory of Metaponto, historical circumstances and chance have combined to leave us not only one of the best preserved classical landscapes in the Mediterranean, but also a document of enormous value to help us interpret it.

The Heraklea Bronze Tablets, now in the Naples Museum, have been known by scholars since 1732, when they were recovered from the Cavone River on the southern edge of the territory of Metaponto. They recorded the official acts of a commission of the adjacent colony of Heraklea (founded in 432 B.C. by colonists from Taras) entrusted with restoring the property lines and revenues of the sacred lands of Dionysos and Athena Polias. Tablet I contains a report of the commission (94 lines), a statement of regulations (84 lines) and a list of renters, their guarantors, and the rent (eight lines). Tablet II is unfortunately incomplete. The preserved text of both tablets, however, is without gaps. The language is technical and in a Western Greek, Dorian dialect, containing some unusual words, for which the only possible translations are intelligent guesses. As they go about their business, however, the commissioners describe the farmland in its essential features. There is nothing extraneous, nothing out of place. The occasional vivid detail and the panoramic view combine to create the effect of a tapestry. Until recently though, it was, like many documents of economic history, separated in space as well as time from its corresponding reality—a text without an archaelogical context.

On historical and linguistic grounds the document belongs either to years just after the expedition of Alexander the Molossian, around 330 B.C., or to that of Pyrrhus of Epirus, who won the first of his costly victories over the Romans at Heraklea in 280 B.C. Thus it dates to either the beginning or the end of the period of maximum settlement in the territory of Metaponto, and by happy coincidence, to that of the accumulation of organic material in the collecting basin at Pizzica-Pantanello.

Its date, the proximity to Metaponto of the landscape described, and its factual reliability make it an interpretive tool of exceptional value for this project. It gives us concrete information about the size of farms in this area of the Greek world, the units in which they were measured, and the means, the boundary roads, by which the boundaries,—"division lines"—between properties were made. It reveals, further, the major agricultural crops and the relative proportions of crop land and woodland on a farm. Finally, it speaks to the perennial and fundamental question of the relation of the individual to the state, of freedom versus control in the economic sphere. [The partial translation which follows, the first to my knowledge in English, is based on the excellent studies of Sartori, Uggazoni, and Ghinotti.]

The survey crew worked during the fall months of 1982 to complete the forty square kilometer area (and the Pizzica survey). Survey leader, Cesare D'Annibale takes notes on site 318 as crew members Ann Patterson, Jennifer Brehob (Sandra Langston and Pontus Riemers, out of camera range) collect material from the surface which will give clues to the nature of the site and its date. V

Heraklea Tablets

Ephor: Aristarchos the son of Heraklidas. Month: Apellaios. The City and the Boundary Inspectors [five names follow] to Dionysos

The boundary inspectors *(horistai)* elected for the sacred lands of Dionysos [same five names] recorded how they set boundaries and metes, measured together, and subdivided lots as the citizens of Heraklea in full assembly decreed.

We measured together, beginning from the boundary road *(antomos)* leading above Pandosia and separating the sacred lands from the private land up to the boundary road marking the division between the lands of Dionysos and that which Coneas, the son of Dion, possesses, and (of that) we made four lots.

The first lot extends in width from the boundary road which flanks the property of Heroideias to the 30 foot wide road which crosses the sacred lands, and in length from the area of the springs *(aporoa)*, [or perhaps "silted up" area] above to the Aciris River. In this lot were measured 201 *schoinoi* [approximately 234,105 square meters based on a measurement of 11,105 square meters for a *schoinos*] of arable land *(erregeia)*, 646 and a half *schoinoi* of scrub *(skiros)*, uncultivated land *(arrektos)*, and woods *(drumos)*.

[Three more lots are similarly described]

The total of all land we recovered for Dionysos was 738 and one half *schoinoi*. We recovered it, instituting legal procedures of thirty days against all of those making sacred land into private land. This same land, as much as we recovered, was then rented for life for 300 *medimnoi* a year, and all the lands of Dionysos for 410 *medimnoi* and one *caddix* a year. [One *caddix* is equivalent to 1/72 of a *medimnos;* the latter approximately equivalent to 52.50 liters (1.49 standard bushels or 89.39 pounds). The payment was made in barley.]

We also set up boundary stones *(horoi)* on the upper side: one on the boundary road towards Pandosia, which delimits Heroideias' land and marks the boundary between the sacred land and private land, having moved it from [the area of the] springs onto private land so that it will not disappear, submerged by the muck, like earlier boundary stones. Another, we placed on the boundary road which flanks the land of Phintias, near the papyrus grove and the canal, having likewise moved it onto private land. Other countermarkers *(antoroi)* we placed opposite these along the carriage road *(amaksitos hodos)*, which leads through the sandbank *(kharados)*, beside the forest, the slabs *(stalai)* in the sacred ground, the countermarkers in the private land, leaving a boundary road of twenty feet . . . those boundary stones which face on the sacred side of the boundary road (are) inscribed "Sacred of the lands of Dionysos" *(hiaros Dionuso khoron);* those in the private land, "countermarkers". Likewise, we placed two intermediate boundary stones *(messoroi)* also on the boundary road which flanks the property of Phintias, two on the road which leads from the City and from Pandosia across the sacred lands, and two on the wooded hills near the cheese makers *(tureiai)*.

Contract of Lands of Dionysos

Ephor: Aristion. Month: Appellaios. The Magistrates *(polianomoi)* [two names] and the Boundary Inspectors [the same five names as above] rent the sacred lands of Dionysos, as they are, for life as the Herakleans have decreed.

The renters will enjoy continued usufruct as long as they furnish guarantees and pay the rent yearly on the day before the month of Panamos. Further, if they pay ahead of time, they shall carry it off to the Public Granary *(Damosios Rhogos)* and they shall measure, before the grain inspectors *(sitagertai)* annually placed in charge of the public measure, full measures of clean grain, as fine as the earth produces. . . .

The renter of the first lot, located along the boundary road which leads above Pandosia beside Heroideias' land as far as the thirty foot wide road, shall not plant less than ten *schoinoi* of vines, and in the soil suitable for olives, not less than four olive trees for every *schoinos*.

If any one enters, or pastures his flocks, or carries off something from the sacred land, or cuts a tree, splits or saws it, or damages anything else, the renter will prosecute to the fullest extent and will have himself whatever he should get as compensation.

The ditches running across the lands, the creeks, [the renters] shall not deepen, nor change their course; they shall neither dam up the water, nor divert it away. They shall clean up, as often as is necessary, the streams running beside their land. They shall not plow the roads which have been designated as such, nor block them or close them to traffic . . . whichever of these acts contrary to the contract they commit, the annual magistrates *(polianomoi)* shall fine them and punish them until they shall comply with the contract. No renter shall cut, split, or saw any tree, nor shall another do it for him. He shall not make piles of earth beyond those already existing, nor holes more numerous than already existing on the same land which has been let on contract. He shall not make tombs in the sacred land, nor permit another to do so. If he does not comply, he will be held accountable for damaging the sacred land. He will also build a house *(oikia)* on these lands, a stall *(boon)*, a granary *(mukhon)* and a shed for storing straw *(akhurios)*. The stall, to measure 22 feet long by 18 feet wide; the straw shed not less than 18 feet in length and 15 feet in width, and the granary, 15 feet in all directions. The renters will provide these completed, roofed, and furnished with doors at the same time they need to have planted the trees. If not, they will be fined six *minae* of silver for the stall, four for the shed, and three for the granary. They shall not sell, cut, or burn the wood in the woods or in the scrub. . . . For the outbuildings *(epoikion)* and for the vines, they [the renters] shall make use of the construction timbers they wish. They shall cut as much brush as is necessary for home needs and make use of the scrub and oak groves, each on his own lot. And as many vines and trees as grow old, those enjoying the use of the land shall replace so that the number is always the same. . . . If by war the renters are prevented from exercising the usufruct, the rent shall be revoked as the Herakleans decide and neither they nor the guarantors shall be responsible for the contractual obligations. . . .

As far as the olive trees and figs are concerned, and all the other fruit trees which exist on this lot [the renter of the fourth lot], shall dig trenches around them, heap up the earth, and perform the necessary pruning . . . [Four renters listed with annual rent in *medimnoi*]

Secretary: Aristodamos, son of Symmachus. Surveyor *(gametras):* Chaireas, son of Damon, citizen of Neapolis [Naples]

Table II

Ephor: Dasimos. The City and the Boundary Inspectors [three names, the same as the first three in Table I]

We measured together, beginning from the boundary road which leads from Caina to the thirty foot wide road which leads to the sea, and from the boundary road bordering to the creek where cattle drink *(boubetis)*, the one that runs from the city across the fields *(guai)*. There resulted an area of seven *guai* [equivalent to approximately 332,500 square meters]. . . . There resulted nine lots of one and a half *guai* from the thirty feet wide road to the (other) thirty foot wide road which crosses the one and a half *guai* lots. . . . [The second tablet is 110 lines long, as compared to 187 for Tablet I, and the last part is missing.]

Fresh plow marks score the limestone cover slab of Tomb 13, one of the more elaborate cist tombs in the Pizzica Pantanello necropolis.

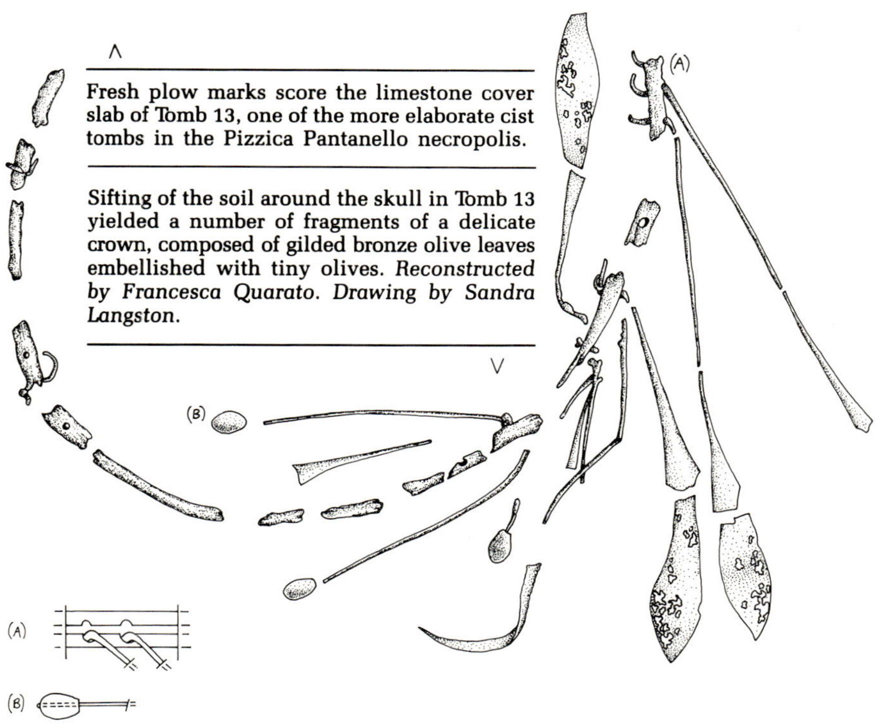

Sifting of the soil around the skull in Tomb 13 yielded a number of fragments of a delicate crown, composed of gilded bronze olive leaves embellished with tiny olives. *Reconstructed by Francesca Quarato. Drawing by Sandra Langston.*

Excavations in the Territory

JOSEPH CARTER

Necropolis at Pizzica-Pantanello, Fall, 1982

The survey is producing the evidence necessary to describe the vicissitudes of the territory and its population over long periods of time. A more intimate view is furnished by the rural necropoleis, such as the one excavated under less than ideal conditions by the survey crew this year. A field, very recently cleared of centuries old olive trees, was being prepared for the winter crop of lettuce when the plow began to scatter the contents of previously untouched tombs. The Superintendency was soon apprized of the situation, and kindly allowed our survey team, back after a month's pause, to assist with the excavation. Fortunately this group included archaeologists, a draftsman, and an architect. By the time the work stopped ten days later, a total of twenty-six tombs had been excavated, drawn, photographed, and put on a general plan. Their contents, including the surviving skeletal material, were catalogued, restored, and stored for future study. This was an exceptionally well-documented salvage excavation. A high percentage of the burials were of children, covered with tiles (*a cappuccina*) and datable to the late fourth or early third century B.C. Adult burials were often primary cremations. One yielded a group of three low denomination Metapontine bronze coins. The typical grave goods, however, consisted of pottery—poor quality black-glazed together with undecorated storage and table wares. An isolated burial of fifth century B.C. date points to a relatively higher level of material prosperity in the early period, though a probable kiln deposit on the edge of the necropolis produced large quantities of good red figured pottery of fourth century B.C. *Continued next page*

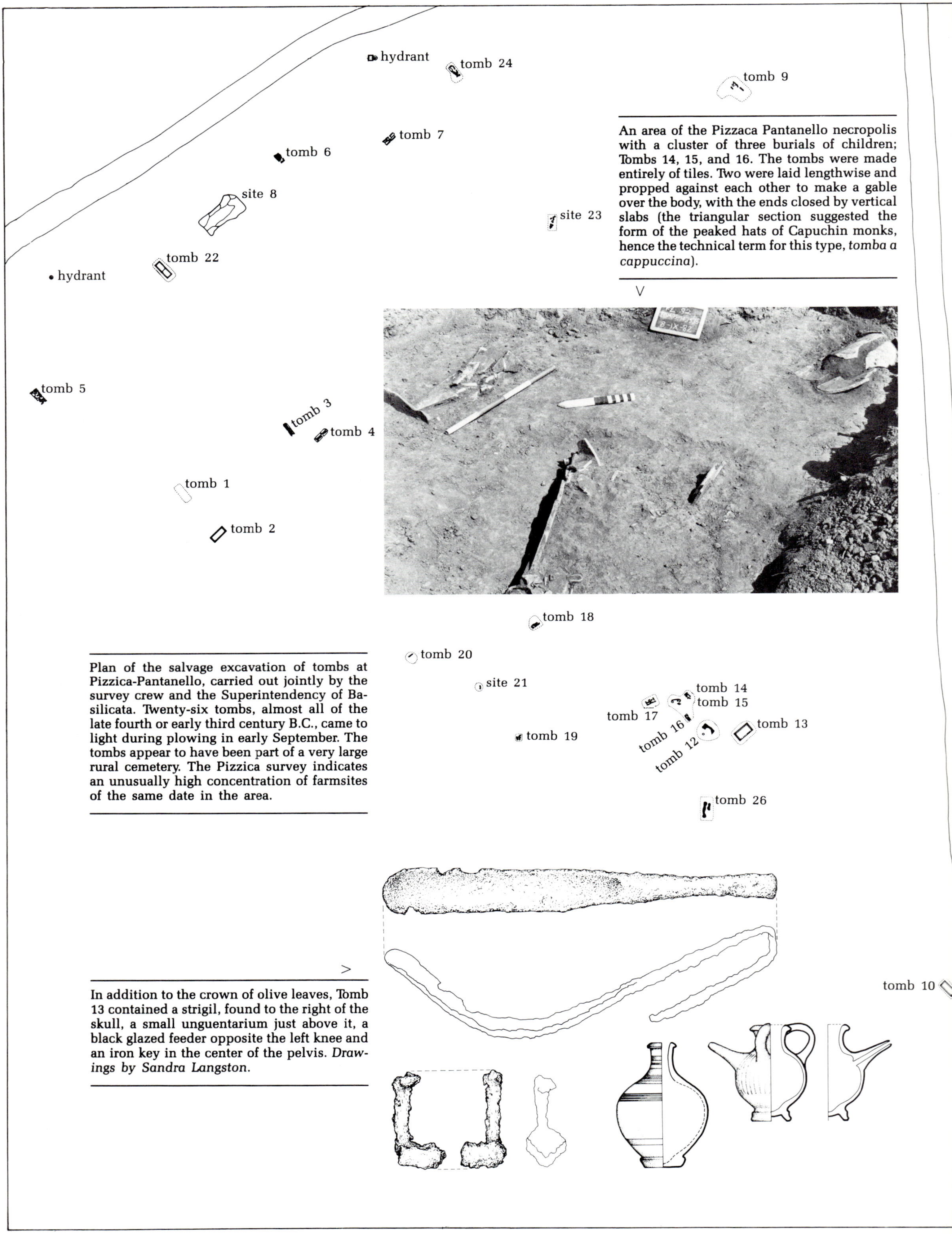

Plan of the salvage excavation of tombs at Pizzica-Pantanello, carried out jointly by the survey crew and the Superintendency of Basilicata. Twenty-six tombs, almost all of the late fourth or early third century B.C., came to light during plowing in early September. The tombs appear to have been part of a very large rural cemetery. The Pizzica survey indicates an unusually high concentration of farmsites of the same date in the area.

An area of the Pizzaca Pantanello necropolis with a cluster of three burials of children; Tombs 14, 15, and 16. The tombs were made entirely of tiles. Two were laid lengthwise and propped against each other to make a gable over the body, with the ends closed by vertical slabs (the triangular section suggested the form of the peaked hats of Capuchin monks, hence the technical term for this type, *tomba a cappuccina*).

V

>

In addition to the crown of olive leaves, Tomb 13 contained a strigil, found to the right of the skull, a small unguentarium just above it, a black glazed feeder opposite the left knee and an iron key in the center of the pelvis. *Drawings by Sandra Langston.*

A partial view of the interior of Tomb 13 with the skeleton of a mature male of impressive stature and grave goods *in situ*. Note especially the absence of hands and the position of the iron key. The scaling rod is marked off in 20 centimeter lengths. The arrow indicates magnetic north. Many of the burials have this same orientation. Others are oriented at right angles; the same direction, approximately, as the division lines of the territory and the majority of burials in the previously excavated rural cemetery at Saldone.

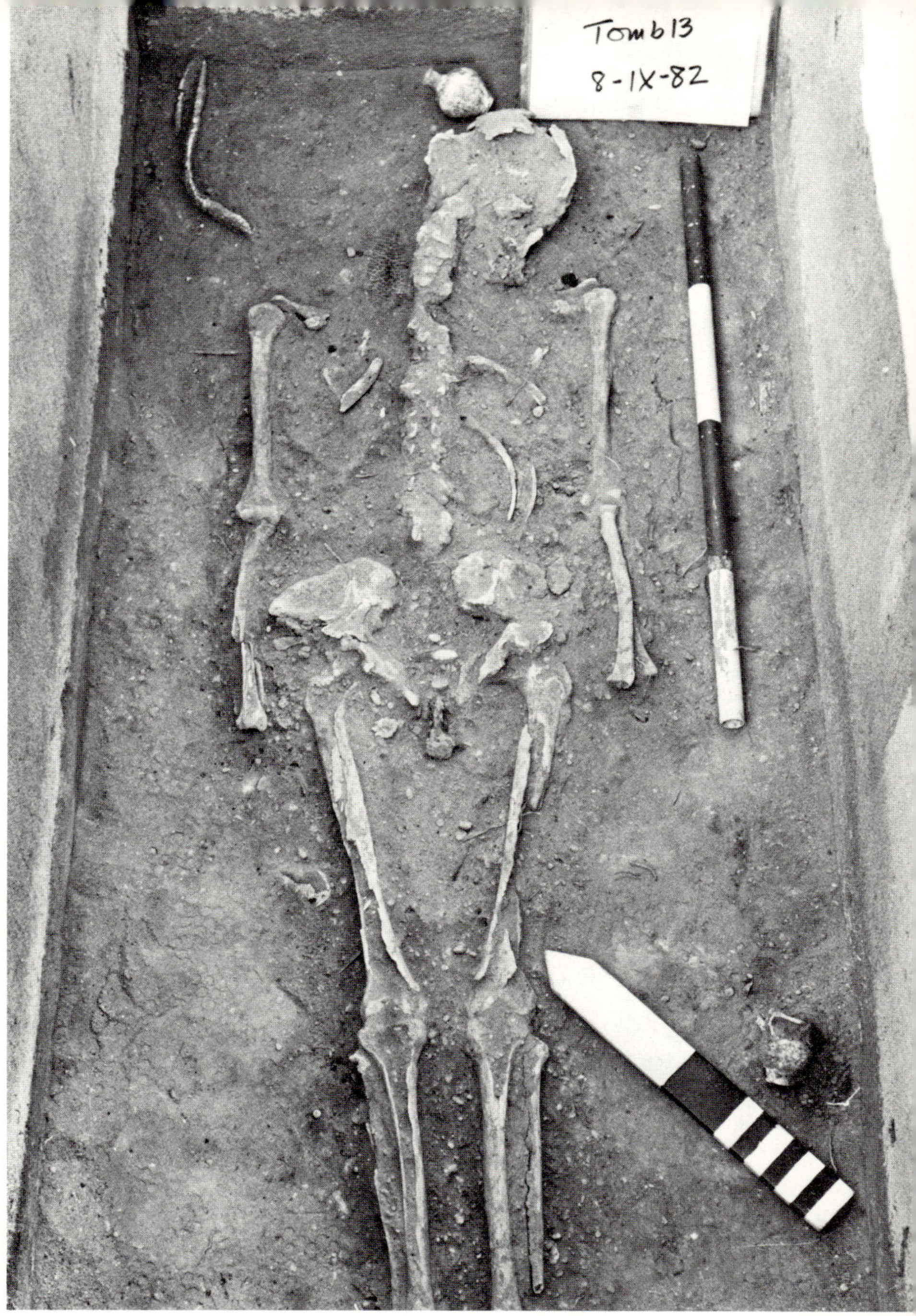

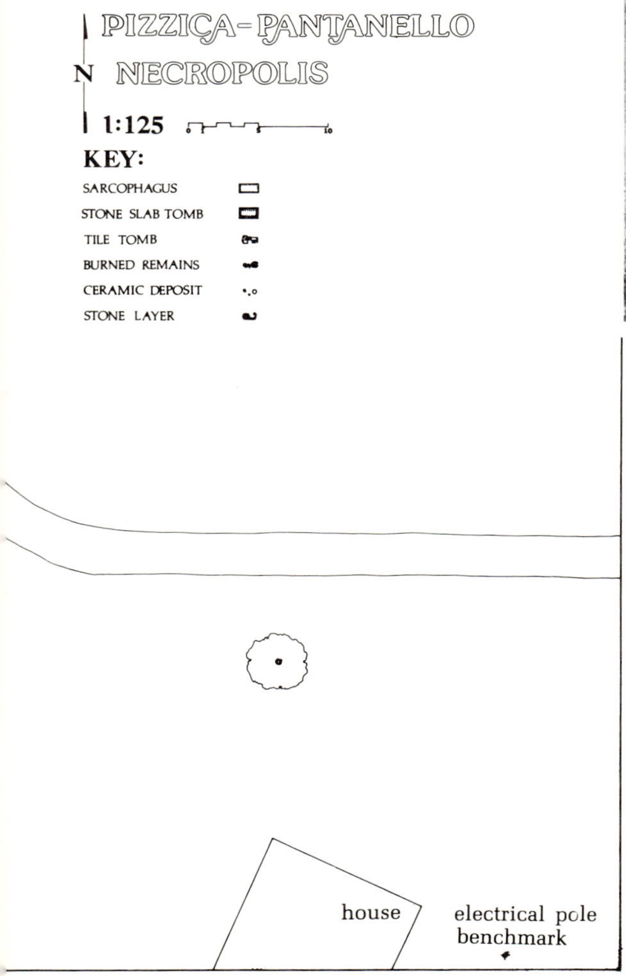

Perhaps the single most interesting discovery was the limestone cist burial of a large, mature male. Near his head were a strigil and unguentarium. He was wearing a very delicate crown of gilded bronze olive leaves. So far this is typical of an upper class South Italian burial. Very much out of the ordinary is the apparent superstitious removal of both his hands and feet, and completely without parallel, to my knowledge, is the iron key placed squarely and deliberately in the center of the pelvis.

The necropolis has been sampled. We hope to continue with a systematic excavation in the summer of 1983.

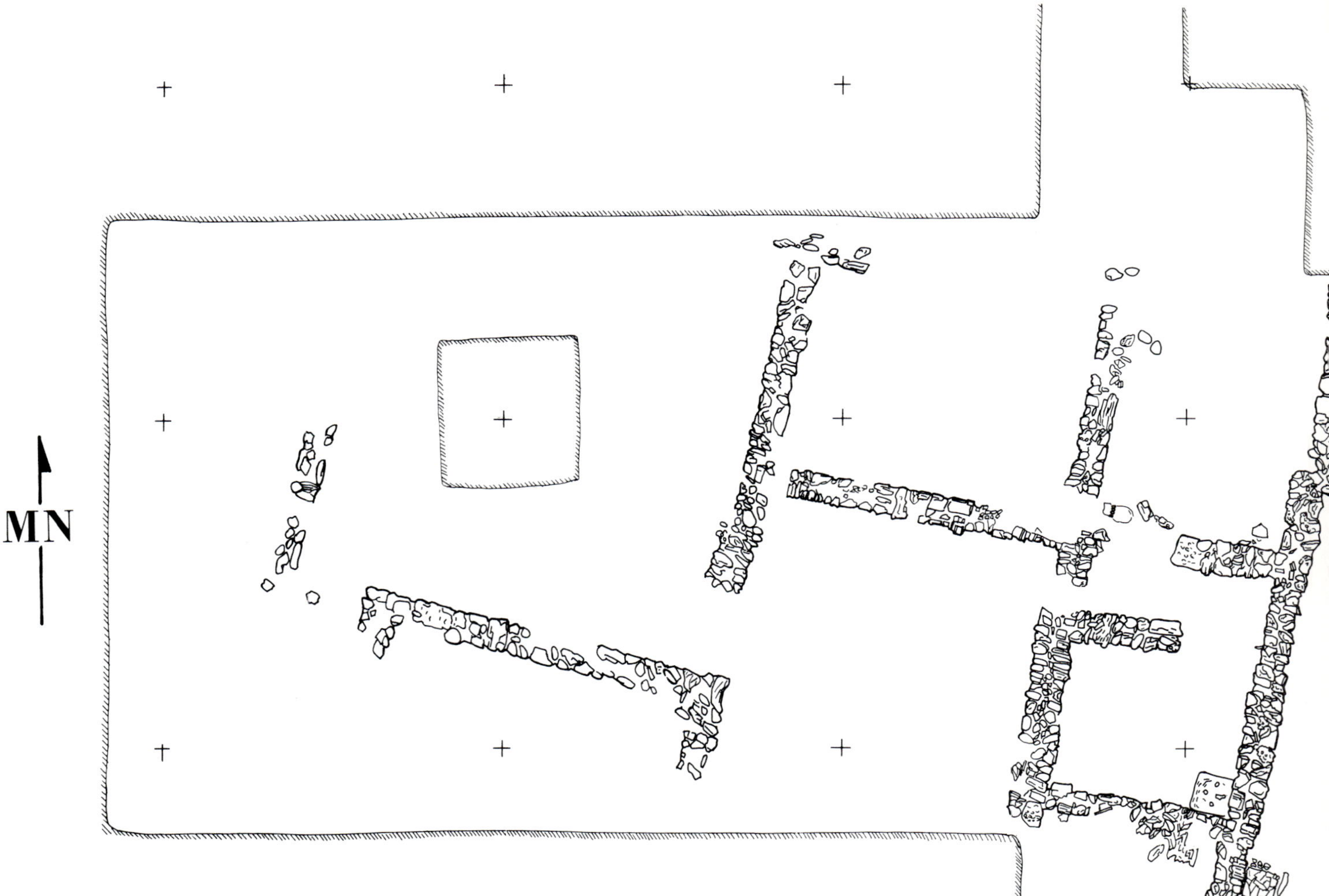

<

Detail of the stone foundations and fallen tile roof of the large farmhouse at Sant'Angelo Grieco. Pottery mixed in and under the tiles indicates a destruction of the final phase of this complex in the first century A.D. Like the farmhouses of all periods in the territory, the stone foundations supported mudbrick walls which, in turn, carried a tile covered wooden roof.

S. ANGELO GRIECO
SITE PLAN SCALE 1:50 0 1 2 3M

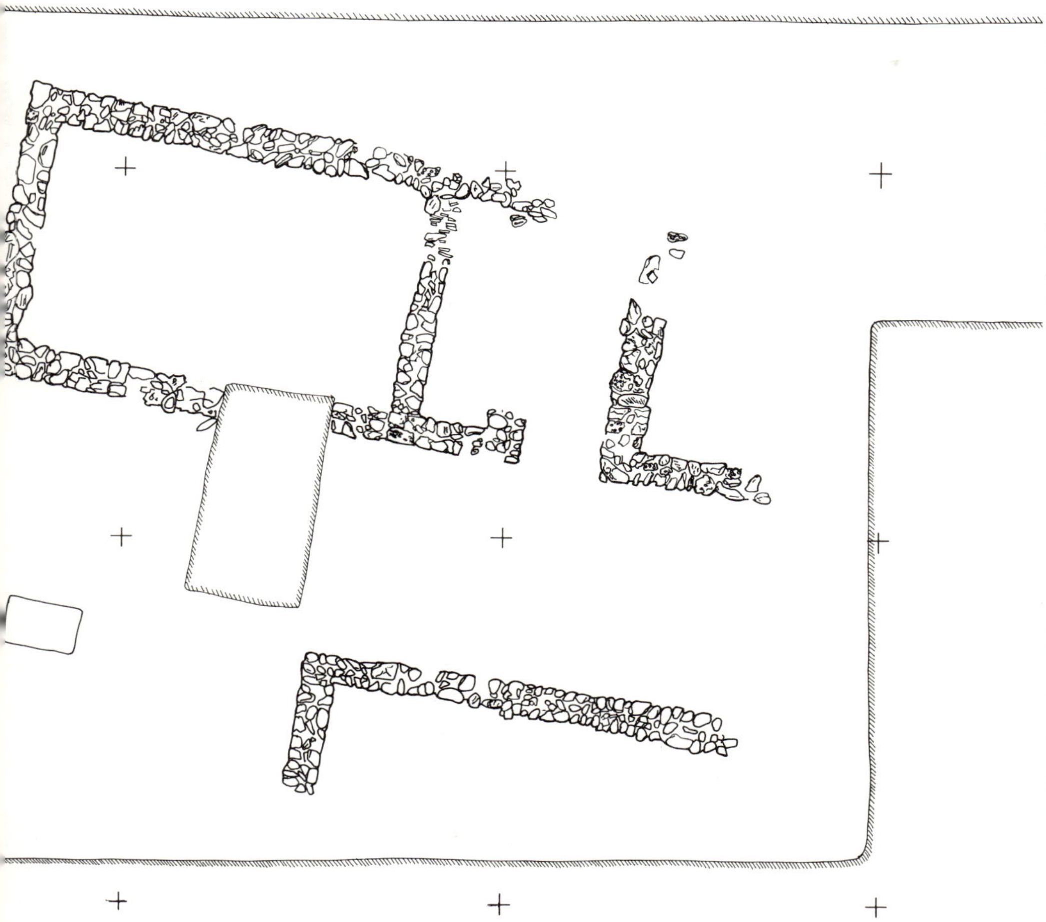

Plan of the incompletely preserved complex of buildings of the Greek and Roman period at Sant'Angelo Grieco. The better preserved foundations in the center of the plan may have belonged to the northwest corner of a building consisting of three wings around a courtyard opened to the south. Note the dog's leg passage between the central and northern rooms of the "west wing." The more fragmentary walls on the western extreme belonged to an abandoned structure of the fourth century B.C.

<

Excavation of the farmhouse complex at Sant'-Angelo Grieco. This was the most extensive of a dozen farmsites in close proximity to an all-weather spring. It had the longest history also. The pottery ranged in date from the late sixth century B.C. to the first century A.D. Earlier buildings, and a nearby Greek sanctuary in the vicinity were robbed of stones for use in the buildings of the Roman period.

V

Farmhouse-Sant'Angelo Grieco, 1981

Farmhouses, of a size to accommodate a single family, located at some distance from one another, were the basic unit of the Metapontine countryside during the Greek colonial phase. This was clear from earlier excavations and surveys; but before our 1980 campaign only one well-preserved example of a large farmhouse, the so-called "Fattoria Stefan" of fourth century B.C. date, and several smaller ones (such as that discovered by the University of Texas team at Pizzica in 1977), had been fully excavated. [See the 1978 and 1980 reports] We have added three more to that number over the last two seasons: Ponte Fabrizio and San Biagio in 1980, and Sant'Angelo Greico in 1981. *Continued next page*

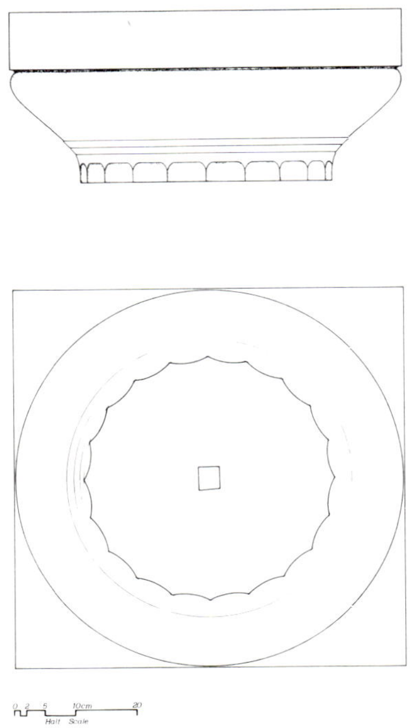

^

Drawing of the Doric Capital of fifth century B.C. date, found in plowed earth in close proximity to preserved walls of the Roman farmhouse. The form is similar to that of the Temple of Hera II ("Poseidon") at Paestum. *Drawing by Deena Berg.*

In 1980 and 1981, our survey crews put a hitherto wholly unexplored area on the archaeological map of the territory. So dense is the cluster of ancient farm sites at Sant'Angelo Grieco—twelve kilometers from the city on the Basento side of the territory—that it ought to be described as a fourth century B.C. agricultural village. Eighteen sites are concentrated in an area of dense olive trees within a radius of 300 meters, centered on an all-weather spring. In a matter of several years irrigation will reach Sant'Angelo and both trees and archaeological sites will likely disappear. With the kind permission of the Superintendency and the land-owner, Sig. Rocco Grieco, and the collaboration of a group from Rice University, soundings were made at four sites in this area. In each case the survey's prediction about their nature and their date was confirmed. Two farmhouses had been almost completely destroyed by plowing. The fourth century B.C. date was confirmed in one case by the discovery of a bronze Metapontine obol. The third site was a tile kiln in a good state of preservation, constructed probably in the fourth century B.C. The final sounding nearest the spring became a full-scale excavation. A single row of olive trees had shielded a portion of the building from the plow. Foundation and fallen roof, as at San Biagio and Ponte Fabrizio, were preserved just below the surface of the field, but deep plowing at a distance of about six meters on either side of the trees had destroyed the rest. This farmhouse, in contrast to the previous two, had clearly defined passageways, and doors with locks, as evidenced by three iron keys recovered during the excavation.

The roof collapsed sometime early in the first century A.D. Grey ware was found, together with stamped Arretine bases, among and under the fallen tiles. A quantity of late sixth and fourth century B.C. ceramics is the major evidence for the earlier buildings on the site. The number and variety of lamps from all three periods is truly remarkable for a rural site.

The fourth century B.C. is well represented by black-glazed pottery, but by far the most interesting of the finds from this site document a period in the life of the territory about which very little is known.

The precise elegant letters of a small fragment of inscription on white marble place it in the fifth century B.C. From the same period is a fine Doric capital, found near one of the walls in an area seriously disturbed by deep plowing. The capital, though of small scale, is comparable to those from the temple of Hera II at Poseidonia, (dated to the mid-fifth century B.C.). Nothing quite like the inscription or the capital has yet turned up in the city of Metaponto itself.

The great temples of the city began to fall into disrepair in the fifth century and were finally abandoned in the fourth. The growing number of discoveries of monumental architecture outside the city, at rural sanctuaries like San Biagio and Pizzica and now at Sant'Angelo Grieco, suggests, as our colleague Dr. Antonio De Siena has observed, that these still only partially explored sites may have been much more imposing architectually in the fifth century than had hitherto been thought. If so, a major shift of wealth toward rural cult places, at the expense of urban sanctuaries, would have taken place at this time.

>

Fragmentary inscription on a thin slab of white island marble, with Greek letters of the fifth century B.C., from the plowed earth immediately over the walls of the Greek and Roman farmhouse site at Sant'Angelo Grieco. Preserved are: [I] first line; I T O [N] second line; [Λ] E Γ third line.

^

The area of the archaic and classical spring-sanctuary at Pizzica-Pantanello at the beginning of the 1982 campaign.

Excavation in the Sanctuary Area, Pizzica-Pantanello, 1982

JOSEPH CARTER

The principal occupation of Metaponto's rural population was agriculture. It is now possible for the first time to describe in very great detail the development of this fundamental aspect of the territory.

In 1978, the excavation of a submerged sacred spring and reservoir at Pizzica produced the first important discovery of ancient seeds at the site. Thanks to the anerobic conditions created by the ground water, flowing as it had in ancient times from the mouth of the spring and into the reservoir, the seeds were in a remarkable state of preservation. The same water, however, made it almost impossible to keep the excavation clean, so that work was suspended until a practical and affordable solution could be found.

One was found in 1982. A "well-point" pumping system was reintroduced, and this time we located it along the very wall through which water filled the reservoir in the fourth century B.C. It is not an easy tool to use, especially in rocky soils like these, and it required constant attention. Much of the success of this campaign was due to the skilled workmen who installed the well-point and kept it running. The entire excavation of eight weeks duration was carried out well below the present level of ground water in the area, and the pump was running 24 hours a day for the whole time.

Continued next page

∧

Workmen installing a "well-point" at Pizzica-Pantanello. A hole must first be dug by ramming the steel tube through twenty feet of sand, gravel, rocks and clay. At the same time water is forced through the tube under high pressure by a "jetting" pump. Once the desired depth is reached the tube is extracted, the yard long, perforated point with its wire mesh filters attached, and the complete "well point" inserted in the hole.

OPPOSITE PAGE TOP >

The collecting basin of the spring-sanctuary as it looked at the end of the 1982 campaign. The well points were set along the northern end of the basin where the water entered. Only by cutting the spring off at the source was it possible to carry out a clean and precise excavation of this area. It happens that the water enters in the same place today as it did when the basin was constructed in the fourth century B.C.

Looking down the line of well points in the collecting basin of the spring-sanctuary at Pizzica Pantanello. The points required twenty-four hour a day maintenance. A small air leak into the collecting tube which connected the points and the small electric vacuum pump, a power failure, or the clogging up of several points was enough to flood the excavation.

OPPOSITE PAGE BOTTOM >

The "well point" proper is a steel tube six meters long with a meter long tip. The heavy pointed tip, perforated for most of its length, is lined with fine wire mesh in order to keep out gravel as water is sucked into the tube while the system is in operation. To install it the point must be driven into the ground, and water under high pressure forced through the tube and out the point. This helps to clear the way. The point will go nowhere, however, without four strong men, a skilled foreman, and lots of patience. The soil here consists of alternating levels of sand, clay, gravel, and larger rocks—an exceedingly difficult combination. It took a week to install fifteen points; the final stage was the construction of a power line to the site for the quiet, small electric vacuum pump. Airtight joints had to be maintained at all times between the points, collecting tube, and the pump. The points needed constant checking to see that they were drawing water and not air. Two days after the pump had been connected, the swamp was dry and excavation could begin. The work proceeded methodically for six weeks.

The relationship between the collecting basin and its unique organic deposits had been documented in all but one vital area, at the north end of the basin, when the water counterattacked. The points stopped drawing and the excavation became a lake. Here the expertise of Giuseppe Di Taranto of the Superintendency (who had first identified the site and played a major role in its early excavation) once again proved invaluable. We also drew once again on the ever helpful Leonardo Torraco of the Azienda Pantanello, who repaired a pump and got it back to the site immediately. Water was forced again through the points under high pressure. The points, cleaned in a little under a half day's frantic work, began to draw water again, and our project could be completed. This was the final challenge in a duel of long-standing with the water at the heart of this sanctuary.

Continued next page

PZ 82
N-8
15 VIII 82
N-8

^

In the foreground: the two channels leading from the spring. It was first frequented in the late seventh or early sixth century B.C. The grill of tiles and the two earthen channels belong to this phase. Over the tiles can be seen two successive phases in the construction of the well. The nicely squared blocks belong to the fourth century B.C., and the uppermost layer of broken irregular stones are of the Roman period.

The main outline of the development of the sanctuary area seemed deceptively clear in 1978. Though this year's excavation focused on the problem of the organic material, it altered appreciably some basic assumptions about the architectural history of the site.

At the center of a complex of structures of later date is the spring, the earliest part of the site to be frequented. Its two channels, cut into the virgin clay and lined by stone, were full of pottery and votive figures of sixth century B.C. date.

This year the "well-point" made it possible to expose the channels completely. They contained besides the pottery, a number of pits of immature olives, pruned vine stems and grape seeds—material evidence, it will be argued, for the offering of first fruits. The channels were cut with care. The eastern branch served as an overflow channel. The western had in its floor beautifully symmetrical pits—shallow ones with votive cups, and one, over a meter deeper, for filling amphorae with the fresh spring water. Some fragments of early sixth century amphorae and enormous cups were found in the fill. The excavation, at this point, was over two meters below the present water table.

The water which filled the channels in archaic times was filtered first through a lattice of carefully laid tiles and river stones. By the fourth century the channels were underground and over the spring rose a well, lined with terracotta drums and buttressed by square limestone blocks. The level was raised again in the Roman period, reusing stones from earlier structures. The worn stepping stone of the last phase and the tile grill of the archaic period encompass over 700 hundred years of rural life.

If the centrality of the spring seems to indicate its placid continuity, the peripheral areas prove that there were major developments in the surrounding countryside. Before the year's work it did not seem likely that the earthen channels could be a part of a truly monumental structure on this part of the site. That view changed when three large squared

blocks of conglomerate were found to the west of the main channel. In section, the relation to two occupation levels (dated with abundant pottery) was clear—an upper paving of river stones, which was laid down after the blocks were in place, and the layer of fine gravel on which they rest. The upper level can be dated to the mid-sixth century B.C. (B2 Type Ionic cups). The earlier one to the end of the seventh century B.C. The conglomerate blocks parallel identical stones in a monumental wall to the east, and both flank the water channels, with the spring at the northern end. The spring in its earliest phase was probably open to the sky, but enclosed at least on two sides with imposing walls. The blocks were quarried nearby and erected on the site less than half a century after the arrival of the first colonists from Achaea.

Various architectual elements from several periods—for example, archaic sima fragments and an entablature, a Hellenistic triglyph and metope block, not to mention the numerous column drums from a large temple or stoa, excavated in 1982—make it almost certain that impressive buildings once stood on the hillside above the spring. The enclosure wall around the spring was, however, the first and the last monumental structure found on this part of the site.

Continued next page

One of the rare examples of imported pottery from the sanctuary area, these fragments belonged to a Middle Corinthian squat oinochoe. They were found in one of the earthen channels of the spring. Early sixth century B.C. Inv.: PZ 81.385P, PZ 82.520P Height: 4.1 cm. Dia (est.): 18 cm.

This small fragment of an archaic Attic red-figure vessel preserves the head of a youthful athlete. It also emerged from the area of the spring, which was frequented throughout the sixth century B.C. This fragment is the latest datable artifact from the area of the channels. Inv. PZ 81.360P Maximum Dimensions: 5.9 × 4.8 cms.

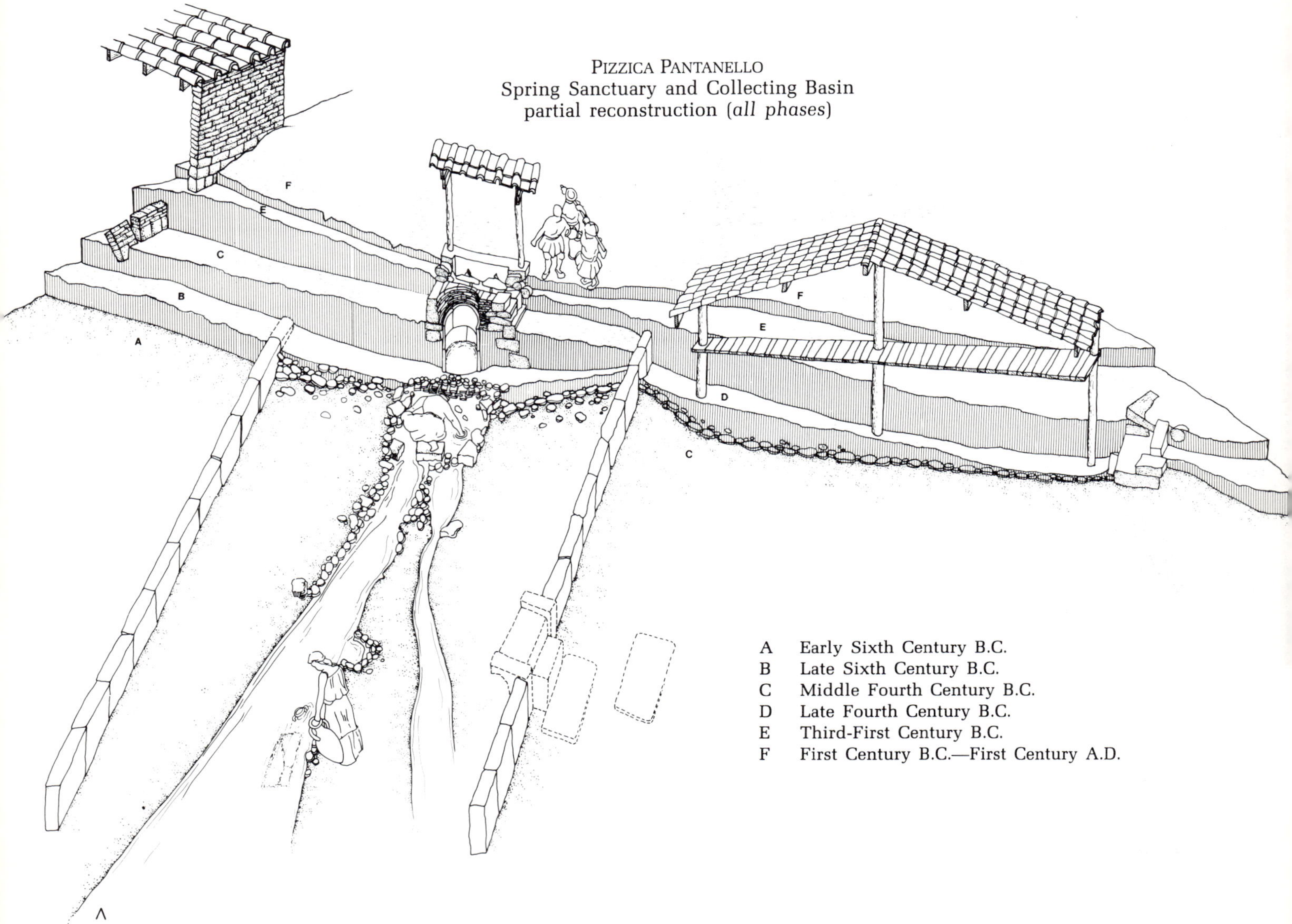

Reconstruction of the spring-sanctuary area at Pizzica Pantanello, with all of its major phases indicated. Left center: the spring and its archaic enclosure walls (eastern wall was reused as the western retaining wall of the fourth century B.C. collecting basin, with steps added). On the right, the collection basin with the late fourth (or early third) century B.C. tile roofed "stall" building, which was probably contemporaneous with one of the phases of the farmhouse on the extreme left. The well replaced the spring (left center) in the fourth century B.C., after the archaic spring had been completely filled in. It was patched up in Roman times. *Drawing by Ann Patterson with the collaboration of Sandra Langston.*

As a result of this discovery, the collecting basin-reservoir to the east had now to be reconsidered. The monumental wall, which we thought original with a covered spring house of the fourth century B.C., in our earlier reconstruction [See 1978 report], had instead been simply reused. Two blocks were removed from the archaic wall and steps of squared blocks of limestone added in the mid-fourth century B.C. to reach the river stone pavement of the reservoir.

The excavation of 1982 revealed the full dimensions of this fourth century reservoir. To the south it was bounded by a dike of impermeable clay. The northern wall was constructed of flat sandstone slabs and riverstones and in a fashion analogous to the archaic grill. The hillside was held back, but water filtered freely into the basin. The run off this time, however, was to the east.

It was possible at last, thanks to the miraculous pump, to clarify the relation between the pavement and the remains of a fallen tile roof which had prompted our earlier reconstruction. The roof belonged to a later structure on the site—in all likelihood a simple shed, perhaps a stall belonging to an early third century farmhouse which stood for a brief while to the west of the spring. The pavement of the basin, on the other hand, can be securely dated to the last half of the fourth century—on the basis of vast quantities of good quality red-figured pottery and terracotta figurines.

^

A general view across the collecting basin from east to west. In the left foreground: the stone reinforced clay dike which closed the collecting basin to the south. Middle ground: the east enclosure wall, built of conglomerate in the archaic period, with its fourth century limestone stairs. Top: excavation of the western end of the sanctuary area.

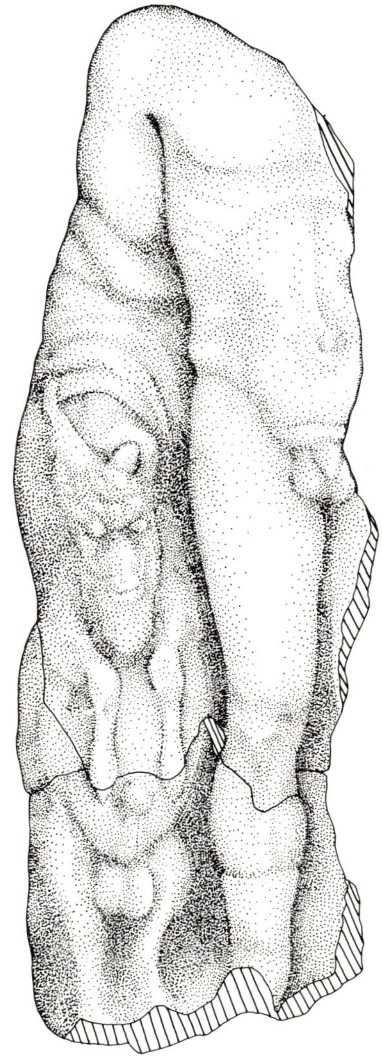

>

Fragmentary terracotta statuette of nude ephebe or hero holding a billy goat by one horn. It was found on the riverstone pavement of the collecting basin and dates to the mid-fourth century B.C. *Drawing by Sandra Langston.* Inv.: PZ 82.232T Height: 21.2 cm.

Among them is a fragmentary torso, perhaps of the tired Herakles, based on the Lysippan model-the earliest extant version of his famous composition? Completely original, by contrast, is the fine torso of a god or hero holding an obscene billy goat by one horn.

Few sites, I believe, demonstrate better than Pizzica the beneficial results of a number of relatively brief campaigns, interrupted by periods of study and reflection. For one, it is necessary with this approach to leave substantial baulks and reserved areas. Thus we could return to the reservoir after a pause of five years and excavate extensive north-south and east-west baulks with the far greater clarity made possible by the "well-point". The results of previous campaigns could then be checked against and integrated into a more precise stratigraphical sequence. The pottery gave us the chronological relationships of the reservoir pavement and fallen roof, and told us that the basin filled up in a relatively brief span of time—less than a century. Results, however, of far greater significance have emerged from the excavation of the baulks, not only for Pizzica, but for the history of the development of agriculture in the fourth century B.C.—in the territory as a whole.

Continued next page

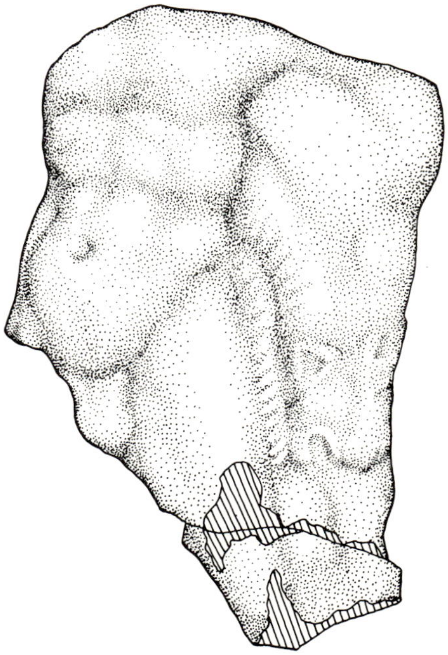

>

Fragmentary terracotta statuette which may represent Herakles with his lionskin, or perhaps a Satyr with leopard skin. This, like the youth with goat, was found on the pavement of the collecting basin, where it was deposited as a votive offering. *Drawing by Sandra Langston.* Inv.: PZ 82.233T Height: 12.8 cm.

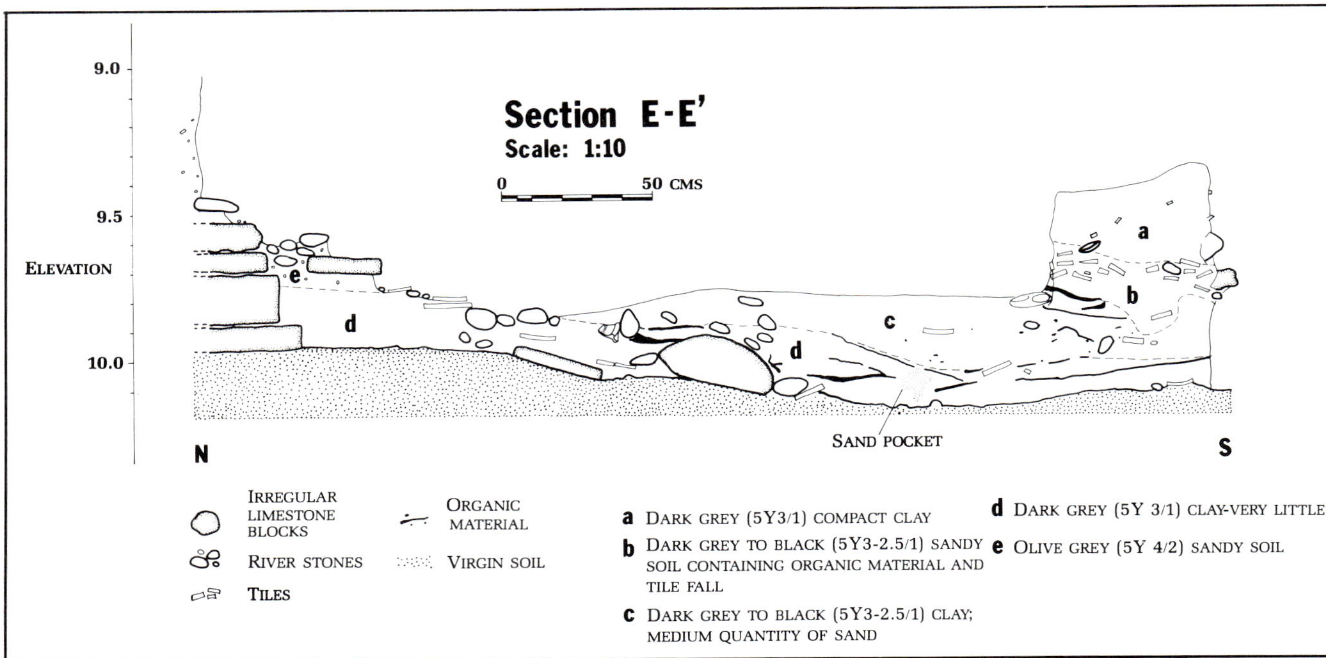

The soil of the baulks and reserved areas was tested systematically for organic material. A thick layer of matted vine, wood, and leaves covered the entire northern end of the basin. It was found directly under the fallen tile of the shed roof. The deposit diminishes to the south where at a slightly higher level there is an isolated and clearly defined layer of earth with many fragments of carbonized wood. At first, soil samples were examined on the spot as they were excavated, using water from the spring, filtered through the well-point, to operate our flotation device. Much of this work was carried out under the direction of paleobotanist Lorenzo Costantini of the Ministero dei Beni Culturali. [His report is presented below.] Hence it was possible to know immediately which areas of the site and levels needed to be sampled more fully. In all, over 5000 pounds of soil were designated in blocks according to position in the stratigraphy of the basin, removed and transported to Costantini's laboratory in Rome.

The material covers a wide range, from plants, like grape and olive, which reflect the agriculture of the territory, to the weeds and aquatic plants which characterize the particular situation of a reservoir gradually silting up.

Preservation in the oxygen free environment of water-logged clay was so good that not only grape seeds, but whole fruit, skins and all, have come to light, together with structures as delicate as insect wings and olive leaves.

Continued next page

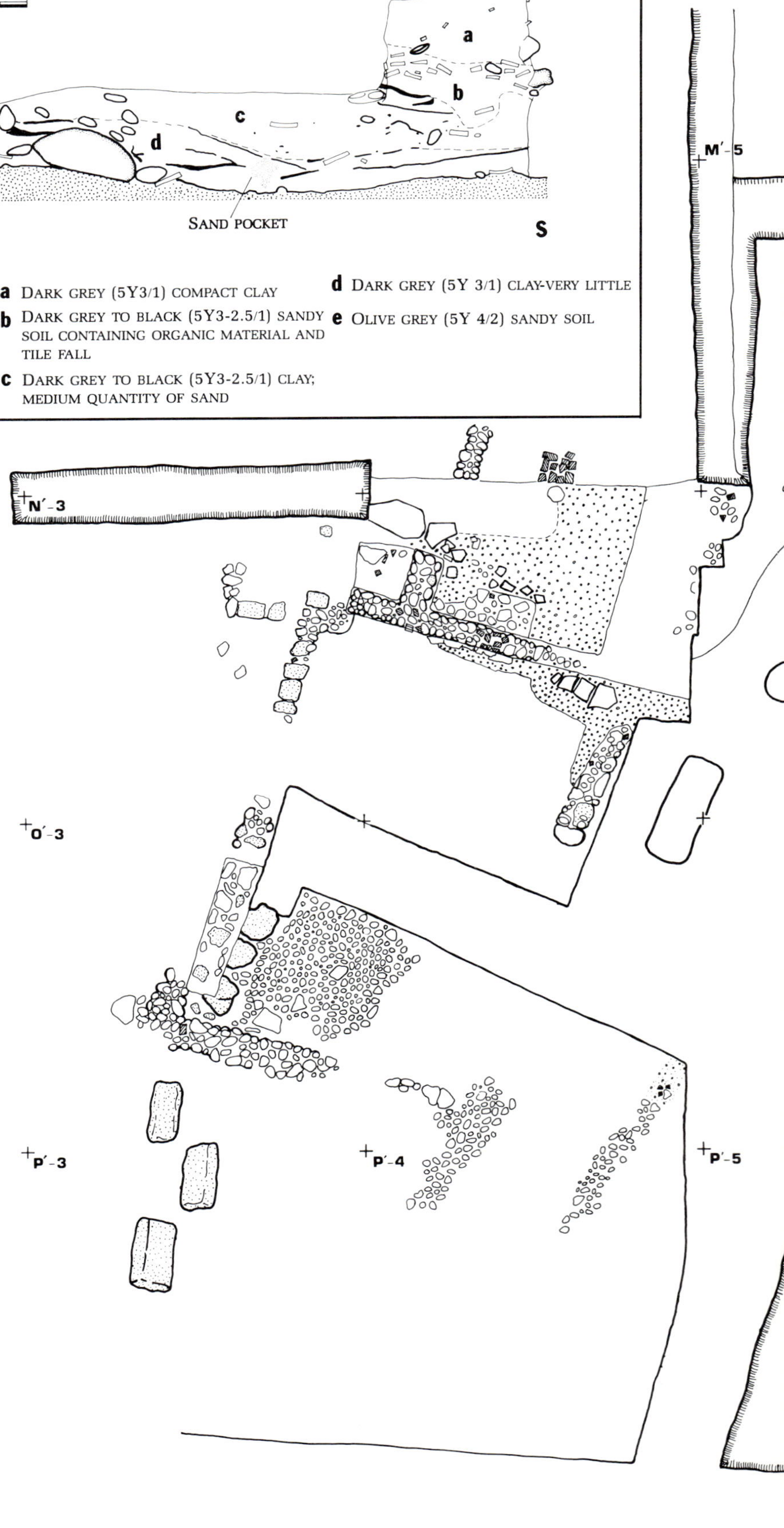

<

A section through the fill at the northern end of the collecting basin (see the plan for precise location of E-E1). To the north (left) the squared sandstone slabs of the northern retaining wall of the basin, to the south (right) the east-west banks (in section). The dark streaks below the level of tiles indicate the concentrations of organic material. *Drawing by Suzanne Jorgensen.*

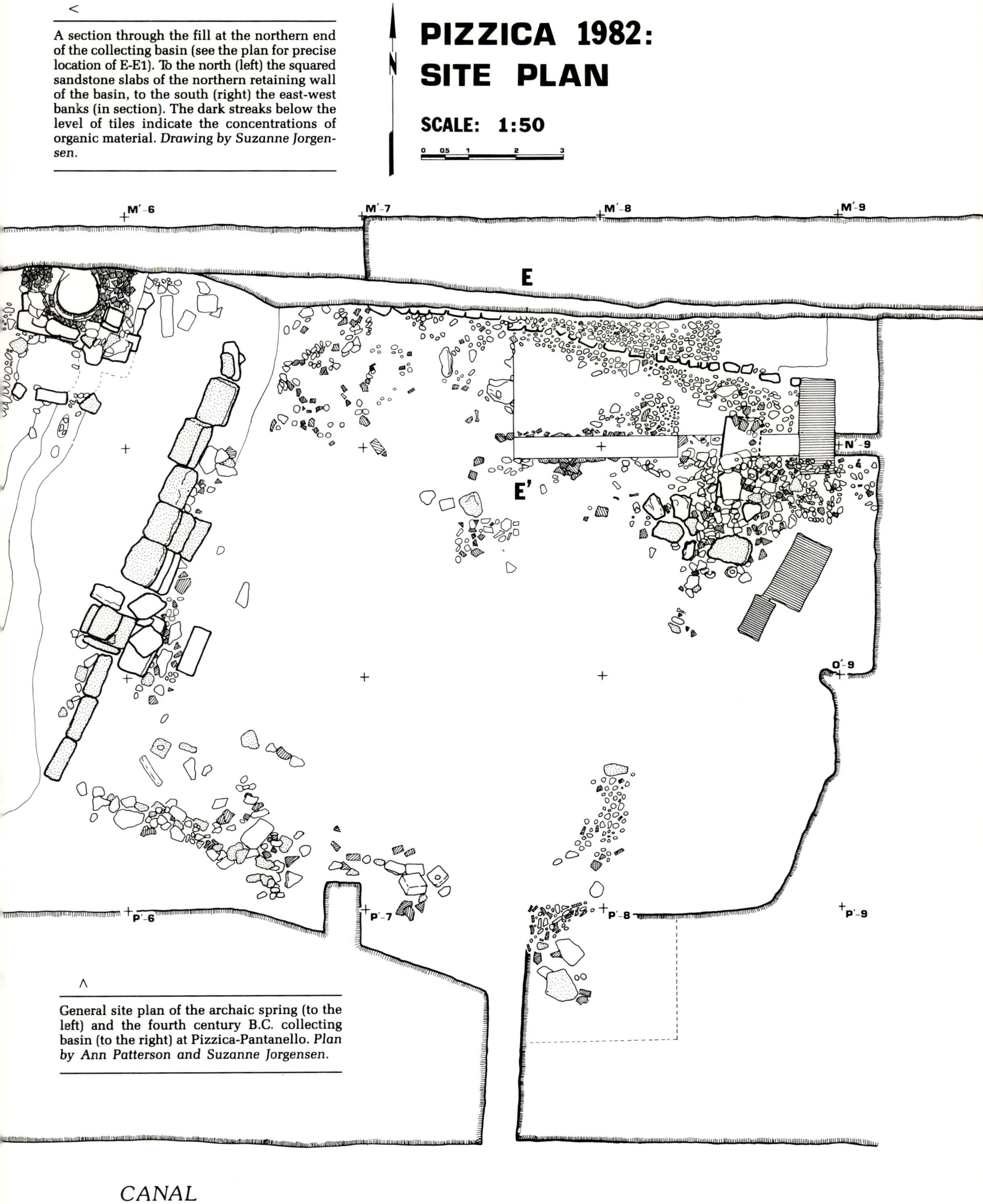

^

General site plan of the archaic spring (to the left) and the fourth century B.C. collecting basin (to the right) at Pizzica-Pantanello. *Plan by Ann Patterson and Suzanne Jorgensen.*

As our workmen opened blocks of organic material in the area of the fallen roof, they were greeted with a pungent aroma they identified immediately as that of stall bedding. It is just possible that bacterial spores—associated with urine—may also have been preserved in this anaerobic environment. Similar conditions on a site in Northumberland have produced this kind of evidence for the Hadrianic period. Enough soil has been left to permit an investigation of this possibility—that we had for a brief while captured the smell of the past.

A great variety of wood was recovered from the reservoir, some of it clearly worked, and in both carbonized and uncarbonized states. It is now under analysis by Costantini and his group. [See below.] Fir, in the form of lumber, as well as olive, grape, willow and poplar used to support the vines, have been identified by sectioning and electron microscopy. More surprising is the presence of maple—a wood employed, it seems from Pliny's description, for fine furniture and nothing else. In some cases a black fungus which attacked the wood in ancient times was preserved.

A series of pollen samples taken from levels ranging in date from late seventh century B.C. to the first century A.D. have added greatly to our knowledge of the plants from this region and its surroundings. This study, which is at a very preliminary stage, is being carried out by Don Sullivan of the Geology Department of the University of California at Berkeley [whose report is included here]. It has already confirmed independently, in broad outline, the process of the silting up of the reservoir, which a study of the stratified organic material (see below) and datable artifacts revealed. The analysis of the organic material in the basin has given us a highly detailed picture of a microcosm of the territory—a particularly important one—which the pollen analysis complements. The results of pollen analysis have significance for the broader picture, and will hopefully begin to document the major changes throughout the thousand year history of the territory. A further preliminary result which may prove to be significant is the presence in archaic levels of *plantago*—a possible indicator that fields were being overgrown and used for grazing in the sixth century B.C. This would be a surprising result, since Metaponto's prosperity, which was assumed to have been based on agriculture, was presumably at a high point in this period.

A major transformation of the Metapontine economy from agricultural to pastoral under the Roman domination, has been postulated on the basis of our faunal studies (see below), as well as on ancient literary sources. Pollen analysis may be helpful here also, though as yet the evidence has not been as well preserved as it has for the earlier periods.

A Note on the Red-Figured Pottery from Pizzica

NANCY JIRCIK

Since the first campaign in 1974 more than 250 fragments, plus some more complete examples, of red-figured pottery have come to light in the excavation of this complex site. Together they document a wide range of Lucanian and Apulian vase painting. The material was examined carefully in the summer of 1982, joins noted, an attempt made to attribute each of the fragmentary works to a painter, and to assign it a date in accordance with the definitive studies of A. D. Trendall. Some of the fragments are from soil disturbed by gradual erosion and by recent plowing, but many are from contexts where the stratigraphy is clear and this fact adds considerably to their interest in context of the excavation. They are among the most precise indicators of the dates we have for the various areas of the site.

The sacred spring and collecting basin were particularly productive of red-figured pottery, which almost certainly was deposited there as a votive offering to the divinity of the spring. This area has yielded a rare, nearly complete thin-walled lekythos, recomposed from many small fragments. A woman is shown seated on a rock, facing a stylized branch and a rather languid, standing hermaphrodite. She holds an open chest to the left; the youth offers her a phiale. The scooped neckline of the woman's drapery, drawn in broad strokes with the ample breasts outlined by curving lines, the foliate details and the square body of the youth are seen on vases from the potters' quarter of Metaponto recently excavated and published by Italian scholars. The vases have been attributed to an Apulian painter of the third quarter of the fourth century B.C. It has been proposed that a Tarantine potter, close to the circle of the Darius painter, transferred his workshop to Metaponto and began producing Apulian style vases.

A number of other ceramic finds from this area demonstrate affinities with material from Metaponto. Fragments found at Pizzica in 1978 and 1982 joined to form a small owl skyphos. The thick, sketchy lines that give shape to the owl's body are similar to those on a fragment from a kiln deposit in the city.

Three Pizzica fragments joined to form part of a large calyx krater, which depicted a youth, nude except for a clamys, leaning against a tree to the right and facing a partially preserved figure seated on a lion pelt. An empty scabbard for a short sword was placed over the shoulder of the second figure, who may represent a caricatured, pot-bellied Herakles. The figures are separated by a low, footed basin. Many, though not all, of the elements of the design found on the vases can be identified with the Creusa/Dolon workshop. A fragmentary nestoris, whose shape is indigenous rather than Greek, was also found in the collecting basin. It has several points in common with the calyx krater and may have been made by the same hand.

>

A miniature Italiote red-figured owl skyphos from the excavation of the collecting basin. Much painted pottery was discovered in the northwest corner of this structure, where it was deposited as votive offerings to the divinity of the spring. It permits the archaeologist to date the basin in its earliest phase; to the mid-fourth century B.C. Inv.: PZ 82.278P Height: 7.9 cms. Dia (max): 8.6 cm.

∧ TOP

A general view from the southwest to the northeast across the collecting basin. To the right of center is the reinforced northeast corner of the basin, and above it the electric pump of the well-point system. The level of the fill before excavation began in 1977 is indicated by the scarp just behind the collecting tube of the system. At its northern end, about twelve feet of earth overlie the basin and its deposit of organic material. *Photo by Aldo La Capra.*

∧

General view across the collecting basin from the northeast to the southwest corner, during the preparation of the final site plan. To the right: the archaic enclosure wall and later stairs. Far right: the excavation around the western enclosure wall. *Photo by Aldo La Capra.*

Apart from the light it sheds on artistic production at Metaponto and the relation of this rural site to the city, the red-figured pottery helps to interpret the development of the sanctuary at Pizzica. The fourth century, coinciding with the development of the Creusa/Dolon workshop in the Metapontine *kerameikos* (c. 400–375 B.C.), witnessed renewed building activity. A greater number of red-figured vessels were offered at the spring. The nestoris fragment may be seen as an indication of the participation of segments of the now Hellenized indigenous population in worship at a place where prior offerings had been purely Greek in character. The Ornate Apulian fragments are relatively few, but of great value in establishing the chronology of the final phase of the Greek sanctuary, and its associated deposits of organic material. They are additional proof that the filling of the basin and formation of the deposit took place in a brief space of time.

^

Sampling of organic material at the northern end of the collecting basin. Small samples (1000 cubic centimeters) were removed whole from vertical cores for analysis in the laboratory in an attempt to discover just how this deposit was formed. The kinds of seeds found at the different levels reveal much about specific conditions and how they changed over very brief spans of time. A study of the organic material adds much precious detail to the stratigraphic description of the basin fill based on more obvious changes in soil and pottery types.

OPPOSITE PAGE TOP >

Another method of sampling soils for organic matter was conducted on the site using a flotation drum (center right). A steady supply of clean water was obtained from the discharge tube of the well point system. To the left is the "jetting" pump, used for installation, and emergency cleaning of the well points.

Carbonized ancient seeds, charred wood and modern roots being collected in the fine mesh strainer at the bottom of the spillway of the flotation drum. Next they are dried and transported to the laboratory for sorting and analysis.

OPPOSITE PAGE BOTTOM >

Bioarchaelogical Research at Pizzica-Pantanello

LORENZO COSTANTINI

Bioarchaeological research in the Territory of Metaponto made a major advance during the 1982 excavation season with the collection of organic remains in the complex which grew up around the sacred spring at Pizzica-Pantanello. There was a twofold scope to this investigation: (1) to trace, by means of analyses of soil samples taken in small stratigraphical columns, the different phases of activity at this particular location; and (2) to document the ancient ecosystem through the study of all the organic remains with particular attention to plants having a use in agriculture and alimentation. In order to carry out the investigation, all the areas which this and previous campaigns had indicated as the major deposit of organic material were very thoroughly sampled.

The site lies in part beneath the present water table, which inundates much of the archaeological zone (and probably did also in ancient times), creating conditions uniquely suited to the preservation of imposing amounts of non-carbonized biological remains. Further, the material was deposited in a gradual way without any particular disturbance so as to permit the formation of a more or less homogenous stratigraphy. The lower limit of the deposit consists of a pavement of river stones which rests on a level of sterile clay, while the upper limit is represented by a dense stratum of tiles, probably the remains of a collapsed roof [decribed above].

The sampling for the stratigraphical columns was made by cutting, from small designated areas, compact blocks of soil, measuring approximately 20 cm on a side and 15 cm high. All the other samples were taken from the major archaeological levels of the site, whose absolute date had been firmly established by the study of associated ceramic material. The samples, a total of 2,500 kgs (over 5,000 lbs of earth), were "washed" at the site as well as in the laboratory in Rome. Washing was carried out using the "flotation drum" technique, which allows for rapid and manageable treatment of large amounts of soil. The sizable quantity of water at low pressure required for this operation was provided by recycling the water channeled out of the site by the "wellpoint". Thus it was possible simultaneously to dry out the ground and to draw water for flotation from a depth of 4–5 meters below the pavement of the basin with the same piece of equipment. The system had a further advantage—the water was "clean" and could not contaminate the samples with extraneous material.

By means of flotation, numerous seeds of plants were collected—spontaneous and wild plants, as well as cultivated ones. They can be subdivided into six groups. Together they document the variegated aspects of the panorama of plant life in this part of the ancient world. One of these aspects relates specifically to the spring and its immediate surroundings. It has in fact been possible to reconstruct in a general way, using the stratigraphy, three phases in the plant life of this area.

Phase one corresponds to the pavement which has yielded, besides numerous seeds of ranunculus (buttercup family, indicating a humid, but not thoroughly submerged ambience), fig, grape seeds, non-carbonized olive pits, and carbonized grains (*caryopsides*) of wheat and barley. In the second phase, the area became swampy with the formation of a marsh or large pools of shallow, slow moving water, as evidenced by the numerous seeds of coontail *(Cerotaphyllum)*, and horned pond weed *(Zannichellia)*, and other aquatic plants. This phenomenon was responsible for the very compact stratum of organic material which progressively covered the northern part of the basin. In this level there were also numerous seeds of fig and grape as well as uncarbonized olive pits.

Continued next page

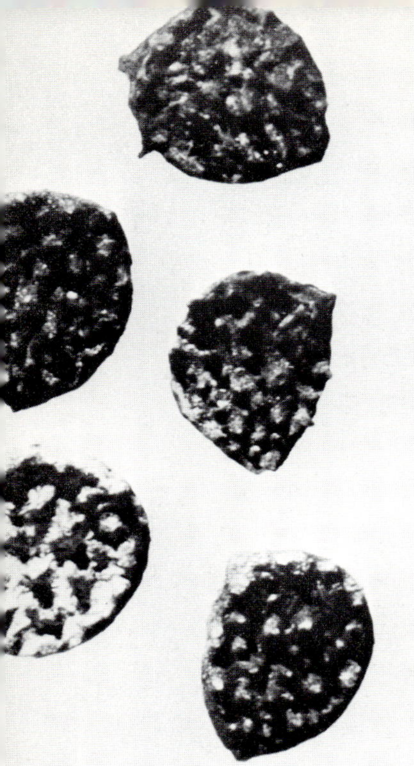

∧ Seeds of Ranuculus (buttercup family)

Fig ∧

∨ Wheat

Grapes: skin, stem and seed ∧

∨ Rachis of wheat

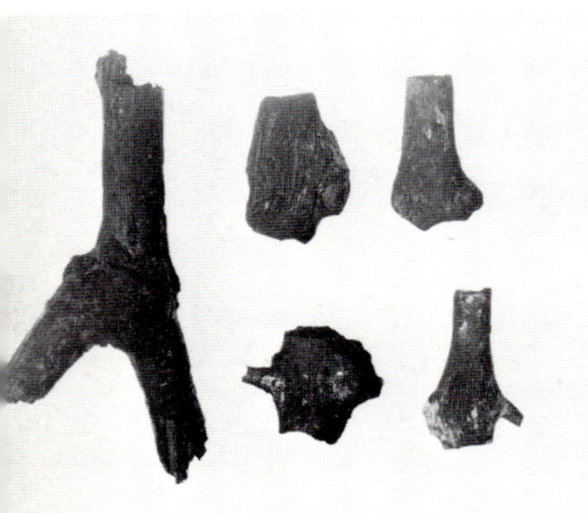

<
Legumes (left to right): chickpeas, horse bean, lentils

Olive mature (carbonized) and immature (uncarbonized)
>

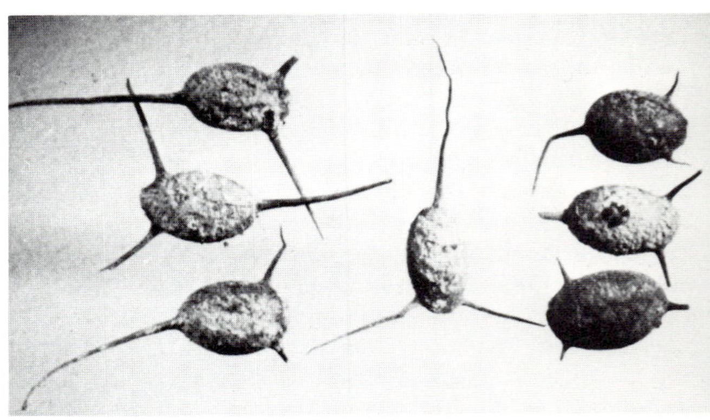

∧ Horned pond weed

Alpha alpha >

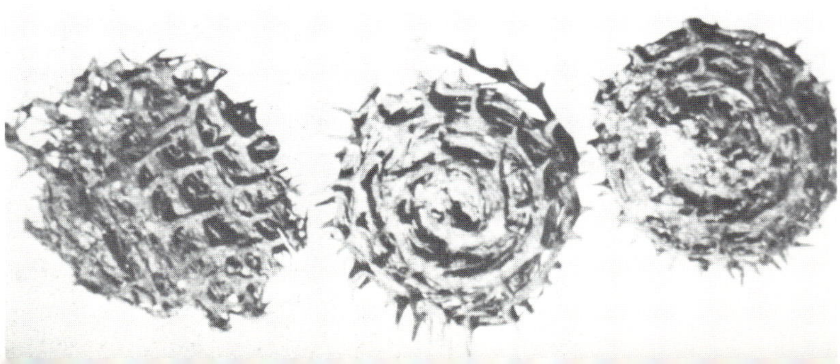

The burial of the marsh under matted organic material and earth—and perhaps a lowering of the water table as well—created new conditions for the use of the spring and basin area. The third major phase, the level immediately over the compact organic material, consisted of loose earth with little clay content, which has furnished us a great number of carbonized seeds. These have been grouped in the following categories:

Cereals The carbonized kernels of cereals belong to two species of wheat, both *Triticum dicoccum* (or *farrum*) and the soft flour wheat *(Triticum aestivum compactum)*, and to one of barley, the six-rowed barley *(Hordeum vulgare)*.

Weeds This group comprizes cultivated plants, like poisonous rye *(Lolium)*, bedstraw *(Galium)*, oats *(Avena)*, and vetchling *(Lathyrus)*, as well as the wild or spontaneous and ruderal plants like blackberry, spurge, and knotweed.

Forage crops To this category in fact only alfalfa can be assigned with certainty, but perhaps the pasture land was also composed of oats and rye grass.

Fruits The presence of olive pits, of fig and grape seeds is constant in all periods. They, like the vines and trees on which they grew (see below) had a great importance for the local economy. (Some carbonized fragments seem to belong to plum pits, but their state does not permit a firm identification).

Comparisons with the Heraklea Tablets

A comparison of these data with the contemporary Heraklea Tablets [see above] makes it possible to understand how fourth century B.C. agriculture was structured, and what importance the various crops had in a plan of pluriennial rotation. From the Tablets it is clear that barley should have been by far the most important crop because the rents are always quoted in terms of measures of barley, while wheat is not mentioned at all. The evidence from excavation at Pizzica-Pantanello—even if the numbers of excavated seeds have a very limited significance—would suggest that wheat was the more important here. In fact, wheat is present in both hulled (suitable for porridge) and naked forms (ideal for bread).

In the definition of the rent contracts for the lands of the sanctuaries of Athena and Dionysos, the modalities governing the cultivation of the grape, olive, and fig, all of which are present at Pizzica-Pantanello in abundance, are laid down. No information, however, is furnished by the Tablets concerning the legumes, which are present at Pizzica in five species. We cannot know, therefore, whether they were a part of the pluriennial rotation as a field crop, or were simply limited to vegetable gardens.

This agriculture appears well defined in all of its aspects, with a differentiation of products into those classes indicative of a highly specialized economy which could satisfy all the needs of a community.

Woods

Together with the analysis of the seeds, a study of the numerous remains of wood preserved by the particular conditions of the earth in the basin was carried out. In addition to the many charred pieces, there were also significant unburned elements of lumber from structures as well as simple cut branches. The woods identified, through sectioning and electron-microscopy, are: fir *(Abies alba)*, maple *(Acer)*, fig *(Ficus carica)*, willow *(Salix)*, and the vine *(Vitis vinifera)* and poplar *(Populum)*. Noteworthy is the presence of a trunk of an olive tree, squared and channeled, and of branches with traces of cutting, which indicates, very probably, the pruning of olives.

The above mentioned analyses were carried out in the Laboratory of Bioarcheology in the Museum of Oriental Art, Rome, with the collaboration of L. Biasini and R. Bucci for the washing of the soil samples and analyses of the seeds, G. Cucinotta for the analysis of the wood, and M. Marzi for drawings of the material. *Continued next page*

Λ

A piece of a fir plank from the collecting basin. *Photo by Lorenzo Costantini.*

Electron scanning microscopic view of a cross-section of vine stem, magnified three hundred times. *Photo by Lorenzo Costantini.*

V

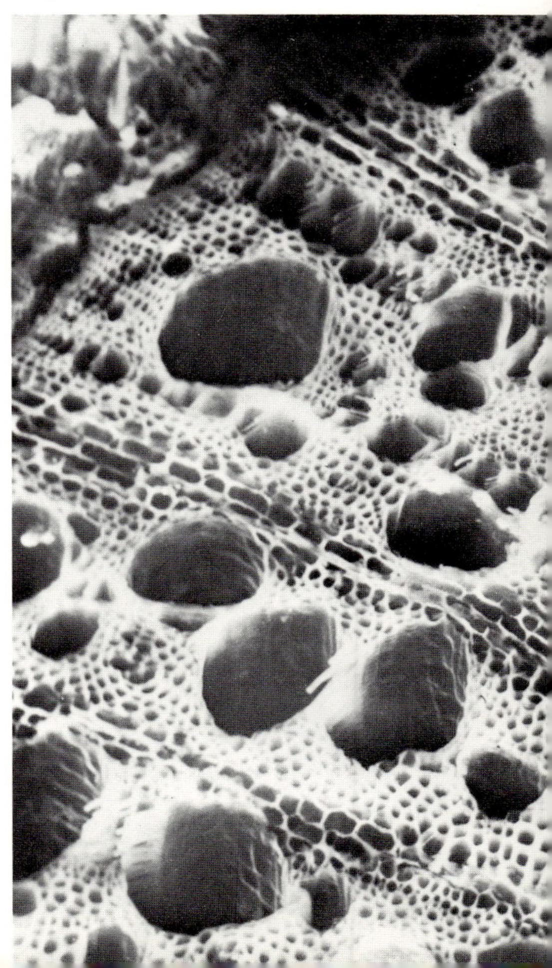

Table 1

Pizzica-Pantanello
Distribution of Seeds
by level in the Area of the Collecting Basin

x-present xx-numerous xxx-abundant	LEVEL I Base level (mid 4th century B.C.)	LEVEL II Compacted organic material (early 3rd	LEVEL III Carbonized organic material century B.C.)
Cereals			
Triticum dicoccum/emmer	x		xx
Tricitum compactum/wheat	x	x	xxx
Hordeum vulgare/barley	x	x	xxx
Legumes			
Cicer arietinum/chick pea			x
Lens culinaris/lentil			x
Pisum sativum/field pea	x		x
Vicia faba/broad bean		x	xx
Vicia ervilia/bitter vetch			x
Forage crops			
Medicago [sp]/alfalfa		xx	x
Avena sativa/oats			x
Lolium temulentum/rye grass	x	x	x
Wild and Spontaneous Plants			
Carex [sp]/sedge family	x	x	
Euphorbia elioscopica/spurge	x	x	x
Galium [sp]/bedstraw		x	x
Lathyrus [sp]/vetchling		x	x
Poligonum [sp]/knotweed		x	x
Ranunculus [sp]/buttercup family	x		
Rubus [sp]/blackberry	x	x	x
Sonchus [sp]/sow thistle	xxx	x	
Cerotaphyllium demersum/coontail	x	xxx	
Zannichellia [sp]/horned pond weed	x	xxx	
Fruits			
Ficus carica/fig	xxx	xx	x
Olea europaea/olive	xxx	xx	x
Vitis vinifera/grape	xxx	xx	x

Preliminary Results of Pollen Analysis Pizzica-Pantanello, 1982–83

DON SULLIVAN

The best evidence of the nature of the vegetation at the Pizzica site comes from pollen samples spanning the period from the middle of the 4th to the early 3rd centuries B.C. (about two thirds of the samples are from this time span). There was considerable variation in the abundance of several important pollen types during this period. The most abundant were Olea europa (olive), Graminae (grasses), Chenopodiaceae-Amaranthaceae (goosefoot-amaranth), Typha ssp. (cattail), Cyperceae (sedge) and two as yet unknown pollen types. Additional reference material from specimens in the University of California herbarium collections are being prepared in an effort to identify the unknown types. One probably represents a member of the rose family.

Of lesser importance were Quercus ssp. (oak), Castanea sp. (chestnut), Compositae (sunflower family) and Alisma-type (arrowleaf) pollen.

The preliminary pollen counts suggest several changes during the middle 4th to early 3rd centuries. Agriculture, including olive and grass production, appears to have been very important locally. The combined olive and grass pollen accounts for 30–40% of all the pollen encountered. There is a temporary drop in olive in the late 4th century and a decrease in grass pollen in the early 3rd century. Vitis (grape) pollen was only rarely encountered in the samples from this period. Vitis is not a prodigious producer of pollen. Nevertheless, the very low counts in this pollen (less than 1%) suggest that it was not an important local crop.

The aquatic pollen types, Cyperacea, Typha and Alisma-type, show increases from the mid-4th to late 4th century. This implies that a marsh surrounded the reservoir, and that this marsh increased in extent during this period. There is at least a suggestion here that the reservoir was getting shallower, either by filling in or by a lowering of the water table. There is also a substantial increase in Chenopodiaceae-Amaranthaceae pollen in the late 4th and early 3rd centuries. This also suggests that the reservoir was getting shallower, with plants from these families colonizing the upper shoreline.

The lack of certain pollen types is also suggestive. Vegetation maps indicate that maquis and mixed coniferous forest should have been located near Metaponto. However, pine/fir percentages are very low (less than 1%) and oak is relatively low (less than 5%). These trees generally produce larger amounts of pollen. Their relatively low pollen percentages may indicate that maquis and coniferous forests were not very widespread near the site. This assertion must be considered cautiously however. It is possible that the vegetation immediately surrounding the reservoir was contributing enough pollen to mask the contribution from outlying areas. Surface samples of the modern pollen rain should shed more light on the interpretation of these percentages.

Samples from the 3rd–1st centuries and from the 6th century were not very productive. But recently it was possible to get good pollen samples from a few of them and data from these samples should be available soon. A cursory look at 6th century samples showed high numbers of Plantago ssp. pollen. High percentages of this pollen type are generally thought to suggest fairly intensive grazing.

Preparation of other samples which had not earlier yielded pollen in sufficient quantity continues and pollen counts for most samples are being increased in order to maximize the data from this material.

A partial view of the collecting basin and well (upper left) in October of 1982 two months after the well points were removed. The water is back at its normal level. Among the aquatic plants in evidence are two (coontail and horned pond weed) whose seeds were also present in the fourth century B.C. The basin would have appeared very much as it does here when it began to go out of use in the fourth century B.C.

V

PZ -81
J'-1/I'-1
Batt. 13B
CP 12Ja-81

Excavation, Pizzica-Pantanello The Deposit of Pottery and Bone, 1980–81

JOSEPH CARTER

The period of Roman domination in Southern Italy, beginning in the third century B.C. saw radical changes in the Greek colonial cities and their territories. The city of Metaponto was reduced to the small area of a *castrum* or *praesidium* by the time of the second Punic War. In the territory outside, as the survey has revealed (see above), in place of the numerous small to medium sized units of the fourth century, there could be found a dramatically reduced number of structures. The terse remarks of ancient writers, like Varro's passing reference to "Metapontine meadows" makes it clear that there was a general depopulation of the territory by the first century B.C. The ceramic factory at Pizzica, which began operation in the second century B.C., was one of the most important rural sites of this period, and the first to have been extensively excavated. The site's character changed radically—from agrarian in the fourth century B.C. to industrial two centuries later, from religious to utilitarian (though, as suggested below by Scali, there is some faunal evidence for continuity in cult practices).

The tile factory on the hill, and a part of the related deposit of pottery and animal bone which cut into the area of the sanctuary were excavated in 1975–76. In 1980–81 the excavation was resumed. The deposit covers an area of about 400 square meters, and varies in depth from two to three meters. By the end of the fourth season, approximately 350 square meters of earth had been excavated, and correspondingly large amounts of pottery and faunal material recovered. Why did this area merit such sustained interest and industry on our part? It was soon realized that the archaeological material from the deposit documented a momentous period for Southern Italy (and the Mediterranean), the last two pre-Christian centuries when the Romans completed the process of assimilation begun already in the third century B.C. This is a period, moreover, for which documents, particularly archaeological ones, are relatively scarce. The Pizzica material consists largely of ceramics and animal bones. The deposit is stratified, albeit in a complex way, and there are enough objects, like coins and stamped Arretine ware to be fairly confident about the dates for all the material.

Much of the pottery was produced locally. There are plenty of misfired vessels, tiles, and other byproducts of pottery works, such as the rings used to separate vases during firing. A particularly attractive example is a seal impression of a gracefully posed Nike in a clay disk.

Continued next page

< OPPOSITE PAGE TOP

Excavation of the kiln deposit of Pizzica-Pantanello, 1981. The deposit is located to the southeast of the Roman tile factory and kilns and northwest of the spring-sanctuary.

A lower level of the kiln deposit. In the middle ground projecting from the trench floor are numerous fragments of amphorae.

< OPPOSITE PAGE BOTTOM

<

This small clay impression made with a seal ring, was probably made for amusement, or as a child's toy by the kiln operator. A lovely, partially draped winged Nike, holding a crown, is the subject. Similar playthings have been found in the kiln deposits of the potters' quarter in the city. Inv.: PZ-81-11OT Height: 3.0 cm.

<

This mould-made "Megarian" bowl, with decoration in three bands, was one of a number found in the deposit, during the 1981 excavation. In the upper register is a frieze of bucrania (cattleskulls); in the center, Nereids on hippocamps; and around the base, Nikes raising tophies. The fabric is greyware; a local product in all probability. Inv.: PZ 81.734P Height: 7.7 cm. Dia (max): 14.0 cm.

A fine series of "Megarian" bowls can be dated to the first half of the second century B.C. Though as yet no moulds for these have been found, the relief is crisp and the motifs (commonly, floral and Dionysiac) though not particularly original are sometimes combined in original ways, as on a hemispherical bowl with bucrania, arimasps, and Nikes with trophies. The quality is often as high, and occasionally higher than that of relief bowls of presumed Tarentine provenience. If the kilns at Pizzica did not produce "Megarian" bowls, they almost certainly were responsible for the enormous quantity of "grey ware" or *pasta grigia* (both the clay and the slip on the vessels are fired a grey color) found there. It is a local variety of the so-called Campana ware—the preferred fine ware of this and other areas of the Mediterranean for several centuries preceding the introduction of Arretine ware, ca. 20 B.C. Our deposit is so large and the forms of the vessels so varied as to suggest that Pizzica may have been a major manufacturing center of this ware in southern Italy.

Utilitarian pottery of a distinctive local variety and amphorae of various types were mixed together in abundance. The fact that the amphorae were found in strata with the more readily datable grey ware at Pizzica is a special advantage in studying that material. It can be dated with some confidence. The amphorae and utilitarian pottery, like the grey ware, have become the object of a special study.

A more fragmentary moulded bowl, in greyware, seen from below. The hemispherical vessel was supported by three spiral feet. The decoration is very delicate acanthus leaves beneath rampant goats flanking craters. Inv.: PZ 81.444P Height: 8.5 cm. Dia. (max): 15.8 cm.

<

>

The pottery shed and laboratory, made available to The University of Texas through the generosity of the Administration of Azienda Pantanello, and the Region of Basilicata, has been the locus of much of our activity throughout nine campaigns. Here the enormous quantities of pottery and bone from the kiln deposit is being washed, sorted, catalogued, mended and studied. Artifacts and bones will then be taken to the magazines of the Superintendency where our draftsmen and photographer will complete the documentation.

The animal bones from the extensive sounding of 1980–81 in the kiln deposit at Pizzica number in the thousands. The importance of this material, however, resides not so much in the unusual size of the sample, as in the fact that the context permits both a relative and an absolute dating of the bones. It is possible to see major changes in the animal populations and consequently in land use on this site. This is no mean result. It may well be that we have a reflection in the faunal record of the transition from an agricultural to a pastoral economy after the Roman conquest, as postulated solely on the bases of written sources by historians like Brunt and Toynbee. If so, there is sufficient evidence from the kiln deposit to trace this gradual transformation and date it with some confidence to the century from 150 to 50 B.C.

A Note on the Amphorae

JENNIFER BREHOB

The Pizzica deposit contained at least 83 amphorae. In many cases only the upper half could be restored. Three, including a Punic amphora, are nearly complete. Many of them are "wasters"—vessels accidentally deformed during firing. No such collection of "wasters" or large kilns of this period have been found in the extensively excavated potter's quarter in the city of Metaponto or at any other site in the territory. Therefore, Pizzica was probably a center for the manufacture of amphorae, as well as common wares, roof tiles, and grey ware.

Continued next page

Washed and sorted pottery on the floor, and restored pieces on the table of the laboratory of Pizzica-Pantanello, 1981.

>

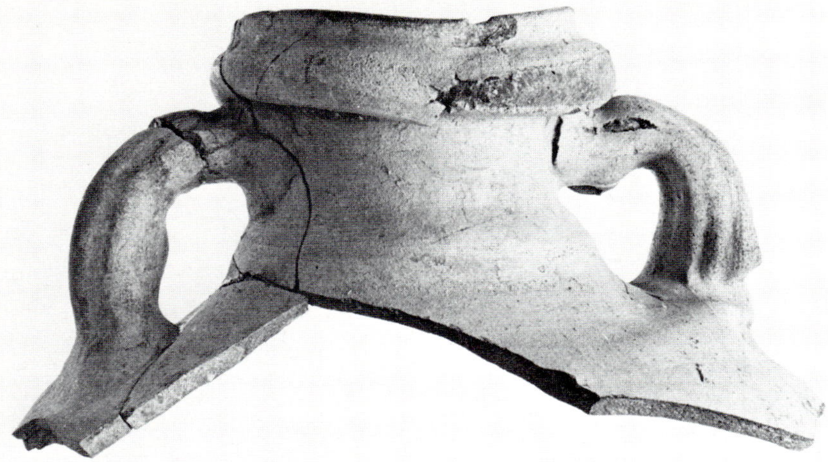

^

A misfired amphora neck from the kiln deposit. This is one of a number of mistakes which prove that the kilns at Pizzica produced amphoras. Inv.: PZ-81-499P Height: 14.5 cm. Diameter (nm): 13 cm.

A nearly complete Punic amphora from the kiln deposit at Pizzica-Pantanello. Inv.: PZ-18-400P Height: 67.0 cm. Diameter (Rim): 16.3 cm. (body): 22 cm.

V

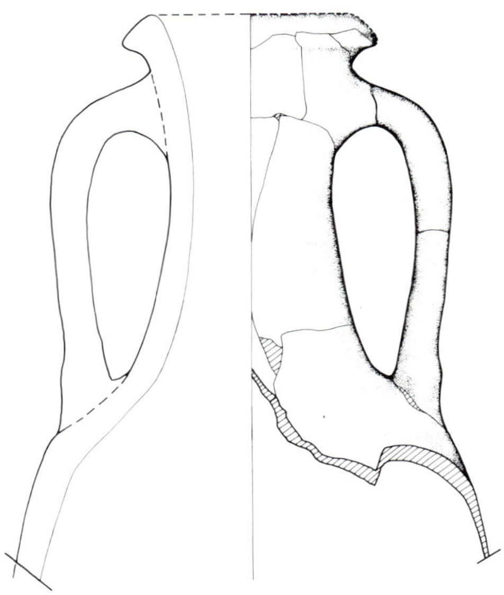

Many of the amphorae have stamped rims or handles. Four different stamps appear on the amphorae from Pizzica. The one occurring most frequently (there are 8 examples) bears the name DAMOKRATES in Greek letters. Only two examples of this stamp have been found in the excavations at Metaponto proper; one on an amphora, and one on a roof tile from the castrum. A goal of this study is to gain a clearer understanding of the role of the factory in the local economy: For what market were the amphorae made? Because there do not seem to be many of the stamped items in the city, it is possible that they were made at Pizzica, filled with goods (preserves, oil, or wine) which had been produced on the fertile land near the kiln, and finally shipped to some point other than the city.

The DAMOKRATES amphorae and the other similarly shaped ones are part of the class called Greco-Italic amphorae. This class was widely distributed in the Mediterranean in the Hellenistic period. Its typology and chronology are currently being investigated by a number of scholars. A study of the Pizzica amphorae—their manufacture, the stamps, and the form—should provide a valuable and timely contribution to this body of research.

<

Apulian amphora from the kiln deposit at Pizzica-Pantanello. *Drawing by Sandra Langston.* Inv.: PZ-81-498P Height: 43.3 cm. Diameter (rim) 12.4 cm.

Stamped amphora handle with the name *Damokrates*, in Greek letters, from the kiln deposit at Pizzica-Pantanello. Inv. PZ: PZ-81-815P

V

Preliminary Faunal Report Kiln Deposit, Pizzica-Pantanello, 1981

BOYCE CABANISS

The Populations The major constituents of the domestic fauna are, not surprisingly, cattle, horse/mules, and sheep/goats. Horses and mules, like sheep and goats, are necessarily taken as a group because of the extreme difficulty of distinguishing between them using skeletal remains. What is unusual and significant about the deposit is the relation over time of the first two groups to the third. As can be seen from Table 1, the quantity of cattle bones and that from the sheep/goat group vary inversely—the cattle and horse/mules are more abundant in the earlier levels and are practically nonexistent later on, while sheep/goats are more abundant in the more recent levels, but are very much rarer in the earlier ceramic deposit. Pigs, another important element of the domestic fauna, seem to be present in fairly consistent numbers throughout the time period (roughly 150 B.C. to 50 A.D.) covered by the deposit. Pigs are the only element of the fauna raised solely as a source of meat. The high percentage of juveniles among the pigs indicates that suckling pig must have been a popular dish.

Dogs are common in several units of the deposit, which is to be expected for a rural area in which stock raising was important. The bones belong to animals of a single breed about the size of a collie or small German Shepherd.

By far the most frequent non-domestic element of the fauna is the red deer. The presence of this big game animal suggests that hunting was an important means of supplementing the diet in this area at least until the beginning of the Imperial period. It also points to the existence of forests in the region large enough to support a population of these woodland animals. Other non-domestic elements of the fauna include rodents, rabbits, and the fox. (It is interesting to note that the fox remains were complete

Continued next page

Λ

Geology graduate student Boyce Cabaniss studies the skeletal material from the kiln deposit, in the laboratory at Azienda Pantanello, 1981.

Bone from the Pizzica-Pantanello kiln deposit. Inv.: PZ 81.237B-9
Cervus Elaphus (red deer): left mandible fragment

V

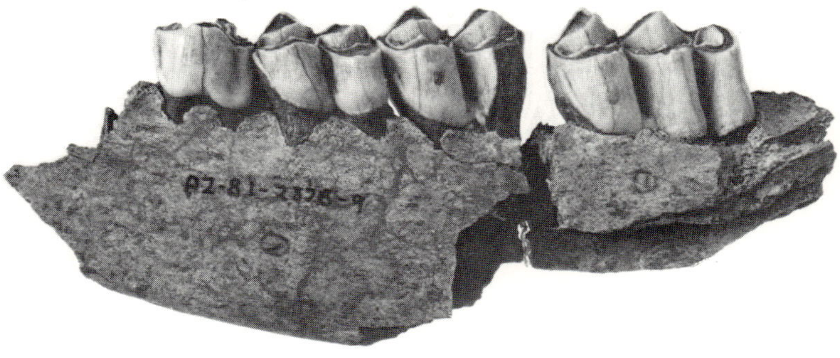

and articulated and showed no signs of butchering. It seems that the ancient inhabitants of this area did not find the fox as tasty a meal as do the modern inhabitants. It is also interesting to note that the fox remains were found very close to those of a chicken. Perhaps he was caught in the act of raiding the chicken house and dispatched by an irate farmer).

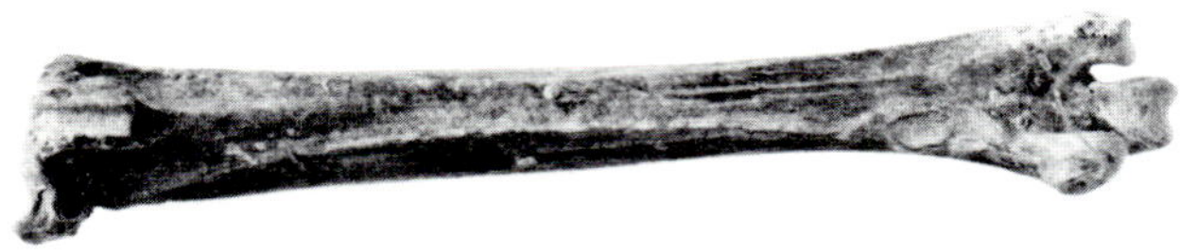

Bone from the Pizzica-Pantanello kiln deposit. Inv.: PZ 81.120B-G
Gallus Domesticus (chicken): right metatarsus

An Interpretation How are we to explain the overall pattern of change in the faunal composition? A closer examination of the remains from the various levels will help to clarify the nature of the deposit. The "ceramic deposit" which sits on virgin soil is the base unit. It is composed of an extremely dense mixture of broken and misfired pottery from the large nearby kilns and large numbers of horse and cow bones. The bones from the ceramic deposit are mostly unbroken. Indeed, 52% of the long bones from cattle and horses are complete. Many of the bones are still articulated so that large parts of the carcasses of the animals must have been thrown into the dump intact. Only one bone from the ceramic deposit shows signs of having been gnawed (despite the presence of dogs). It thus appears that the

Bone from the Pizzica-Pantanello kiln deposit. Inv.: PZ 81.174B-1
Bos Taurus (cow): right mandible of calf

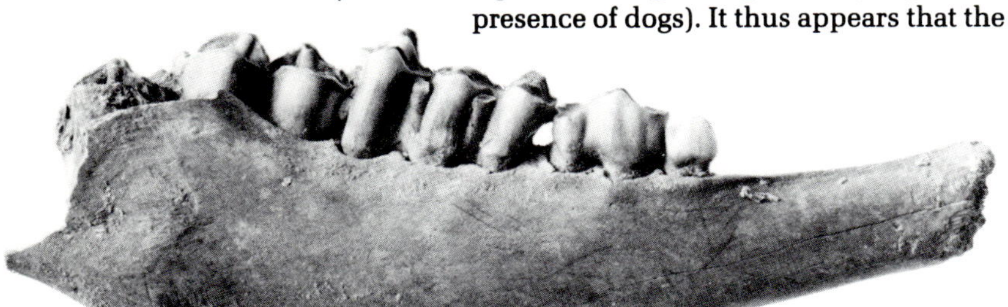

animal remains were covered either by soil or pottery refuse soon after dumping, an obvious procedure given the stench resulting from the deteriorating carcasses. The material also shows no sign of butchering, which is not surprising since there is no evidence that horses or mules were ever eaten, and cattle were eaten only under special circumstances such as after sacrifices. All were too valuable as work animals to serve as sources of meat. On some of the bones there are cut marks, usually appearing on the limb bones or scapulae, which suggest skinning rather than butchering. A set of hyoids from the throat of a cow shows gnash marks, which suggests that the animal was killed by cutting its throat. Sixty percent of the dental material from this deposit came from old individuals, as indicated by excessive wear. Only 4% of the bovid material came from immature animals.

All the evidence here points to the conclusion that these are the remains of old or sick animals which were killed and skinned, and whose carcasses were discarded or used for tallow. This is supported by the fact that all skeletal parts are represented, except for the caudal vertebrae from the tail. The tail may well have been removed with the skin. It is interesting to note that the earliest bone in the deposit coincides exactly with the earliest of the misfired pottery. This site may have been occupied by a small industrial complex which produced, in addition to tiles, pottery, and iron implements, leather and tallow from animals in the immediate environs no longer useful for work. This will serve as a working hypothesis for future study of this material.

The next group of material in order of time, Level 4 (see Table 1), lying down the slope from the ceramic deposit and partially overlapping it, contains quite a number of complete cattle bones and some misfired pottery, but not in such quantities as in the ceramic deposit. It may represent a lateral downslope extension of the ceramic deposit or a later accumulation of material washed down from above. As can be seen from Table 1, deer, pig, and sheep/goat remains be-

Bone from the Pizzica-Pantanello kiln deposit. Inv.: PZ 81.282B-11
Capreolus Capreolus (chamdis): first phalanx

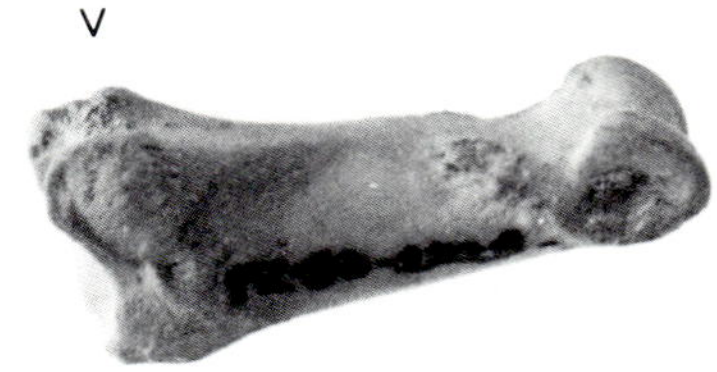

came more prominent in this level. The condition of the material is much different from that of the cattle and horse. None of the long bones of these smaller animals are complete. All are broken and many show the kind of marks caused by chopping, sawing, or other methods of butchering. All of the deer, pig, and sheep/goat remains from the entire deposit are like this: all are scattered and broken and obviously represent kitchen refuse. This is to be expected since pigs and deer were utilized solely as meat sources. The paucity of remains of these in the ceramic deposit probably indicates either that at that time the area was not being used as a kitchen dump or that the cattle and horse carcasses together with the pottery refuse were being deposited in such large amounts as to overshadow the much slower deposition of kitchen wastes.

The deposition of cattle and horse remains declines further in the next highest level and disappears almost entirely above that point, while sheep/goat remains become much more common. This may well have been caused by a slow down and final stoppage of the animal processing operation or by a change in the location of the dump. Since the abundance of misfired pottery also begins to drop off at about the same stratigraphic position, it appears likely that the kilns and animal processing operations both began to decline at roughly the same time. Could the *latifundia* which began to displace the large numbers of small farmers in Southern Italy after the Roman conquest have been responsible? Since most of these large estates concentrated on pastorialism, the cattle and mules used by farmers as draft animals would consequently have become less abundant. The establishment of large scale pastoral operations would of course explain the increased abundance of sheep and goats. The bones from other contemporary sites in the region need to be analyzed to determine if this pattern is repeated over the entire area or if it is merely a local phenomenon.

Table 1

Number of elements of each taxon per level with conservative minimum numbers in parentheses.

<----Early ---->later	Ceramic Deposit 2nd c BC	Level 4	Level 3	Level 2	Level 1 early 1st c AD
Bos Taurus Cow	110(13)	39(6)	29(3)	6(1)	0
Equus sp. Horse, Mule Donkey	55(5)	6(2)	2(2)	2(1)	1(1)
Ovis/Capra Sheep/Goat	13(3)	5(2)	11(2)	36(5)	34(6)
Sus Scrofa Pig	3(2)	8(2)	19(2)	5(2)	6(2)
Cervus elaphus Red Deer	0	3(1)	16(2)	15(4)	10(1)
Canis domesticus Dog	36(4)	0	6(2)	15(4)	0
Gallus domesticus Chicken	0	0	0	2(1)	0
Vulpes vulpes Fox	0	0	0	one complete skeleton	0
Lepus sp Rabbit	0	0	2(1)	0	0
Homo sapiens Man	0	0	0	7(1)	0

Observations on the Faunal Remains from the Territory of Metaponto

SALVATORE SCALI

The following analysis involves a limited amount of faunal material from excavations by the University of Texas at three ancient settlements, Incoronata, Pizzica-Pantanello, and San Biagio. The results are provisional, until they can be integrated with earlier and more recent work.

This research has been aided by the Italian Institute of Human Paleontology, by Cav. Pier F. Cassoli, especially for his generous help with the identifications and Dottor Alessandro Bini for the archaeological bibliography regarding Metaponto.

These three sites, separated from one another by only a few kilometers, existed during three important phases of the occupation of the territory. Incoronata (8th century B.C.) belongs to the precolonial period and the first years after the founding of the agricultural colony by emigrants from Achaean Greece. Pizzica (6th century B.C. to 1st century A.D.) spans the prosperous years of the Greek occupation, the decline of the territory with the Roman domination, and its modest revival during the late Republic and early Empire. Finally, there is the villa at San Biagio, a rare example of the late Empire (late 3rd-early 4th century A.D.). [For the purposes of this report the faunal material from each site is treated as a whole. In fact there is a considerable chronological range in this material—as much as seven centuries at Pizzica—and a variety of specific contexts which affect the interpretation of the material. These matters are dealt with in the full publication of the site, now in progress. Editor's note]

Keeping in mind the selective nature of the material dealt with here, it nonetheless furnishes us with significant elements for precise comparison between the three settlements. These comparisons, in turn, help to illuminate the outstanding characteristics of the successive settlements and their specific activities in the territory—and they provide, at least, some clues to the activities of the territory as a whole in these three periods, and the changes which took place over time.

At Incoronata (8th–6th century B.C.) all the animals typical of a developed animal husbandry are well represented in fairly equal proportions, with a proponderance, nevertheless, towards sheep/goats (36%) and pigs(28%). These are the easiest and most profitable animals to raise, and their presence at Incoronata indicates advanced farming methods probably introduced by the Greeks [as our 1980 report observed]. Cattle and oxen are numerous (19.7%), which are certainly indispensable for tilling the soil. The horse (2.1%) and dog (3.1%) are sparse. There is a noteworthy percentage of deer (7.5%), which no longer exist in this region. This, besides showing that the diet was supplemented by venison, proves, as the ancient writers comment, that the landscape was once more densely wooded. Finally, the aquatic turtle (*Emys orbicularis*) is well enough represented (2.5%) to suggest the presence of swampy conditions and streams favored by the forest cover.

At Pizzica the picture is clearly different. The faunal remains from the site are beyond any doubt the most varied and interesting, and by the same token, the most difficult to interpret.

Immediately obvious is the prevalence of cattle and oxen (36%) which together with horse (14.4%) account for more than half of the finds. This noteworthy increase in bovines may indicate a specialized production. [See above.] The percentage of sheep and goats (10.4%) is sharply reduced in comparison with Incoronata. Deer (9%) maintains its high level and together with boar indicate that the forests had not yet disappeared despite the greatly expanded agriculture of the classical period (see above). There should, however, have existed ample pasture land for the cattle and horses. The quantity of dog bones (8.5%) is unusual, and may perhaps be attributed to the latter's employment in the hunt for boar and deer. In any event, it is clear that large quantities of animals either slaughtered or simply dumped could have provided the necessary requirements for a number of carnivors not strictly indispensible to the human population.

The presence of three other species might likewise seem at first glance either marginal or purely accidental; taken together, however, they are of the greatest interest, and are worthy of special and detailed study. They are the short-horned cow (*Bos taurus brachyceros*, 1.3%), the imperial raven (*Corvus corax*, 0.2%), and the hooded vulture (*Aegypius monachus*, 1.9%). The short-horned cow has been the subject of an exhaustive study by the noted paleontologists, G.A. and A.C. Blanc, who identify it with the Roman sacrificial ox. The breed, clearly distinguished from its larger, long-horned cousins, is well known to students of Roman art. It is represented in the rite of Souvetaurilia, for example, on the Plutei Reliefs of Trajan in the Roman Forum, and in many other reliefs of official Roman art. It is not an accident then that the remains of these smaller oxen were recognized among the bones discovered near the Lapis Niger in the Forum, together with those of the ram and the pig, in the context of a sacrificial deposit of the 5th–4th centuries B.C. The traditional sacrifice and presumably the breeding of these animals for that end persisted through the centuries. Here at Pizzica, we find the remains of these animals in an area where cattle were

^

General view of the canal area at Pizzica Pantanello. Extreme left, middle ground: the south end of the kiln deposit. Foreground: area of late sixth century sanctuary excavated 1974-1976, now overgrown. Center: the excavation of the spring and collecting basin, begun in 1977, being completed in 1982.

raised during the Roman period, as has been established by the study of the deposit excavated in 1980 and 1981.

In support of the identification of the short-horned cattle found at Pizzica as the Roman sacrificial ox, is a second, even more astounding discovery—one which makes the parallel with the situation in the Roman Forum even more convincing. Seven bones of a hooded vulture were found in the vicinity of the sacred spring at Pizzica. This is a very rare bird indeed, and references to it in archaeological reports are practically nonexistent. A noticeable exception are three bones found in the excavation of the votive deposit at the Lapis Niger in the Roman Forum. This is precisely the "big bird" associated there with the bones of the short-horned sacrificial ox.

In their study of the Roman material, the Blancs discuss the characteristics of this rare bird, difficult to capture, and perfectly inedible, but sacred "because it was the least injurious of animals, never touching the seeds men plant or the animals they raise, since it feeds on carcasses and never kills or wounds any living creature." The vulture is associated not only with Roman rites, but occurs also in Etruscan, Greek, and Oriental religious traditions.

It has to be admitted that in spite of the fact that all the traditional animals of sacrifice were to be found at the site, there is no conclusive evidence that a sacrifice was ever carried out. All, except the vulture and the short-horned ox, are quite common. But how can the presence of these two be explained otherwise? The royal raven, finally, as inedible as the vulture, is also necessarily an import at Pizzica, as his natural habitat is in the mountains. He may have the same significance as a bird of omen. Pizzica was a cult place as early as the 6th century B.C. The possibility that it maintained its sacred character despite changes in the local population with the decline of the Greek city and the coming of the Romans, should not be ignored.

At San Biagio, the last of the three sites included in this survey, we return to a more traditional sort of animal husbandry. Around this small but well outfitted farm, a very occasional horse grazed (0.7%) and the oxen were probably just sufficient (8.2%) to work the fields. Milk and meat were provided by a balanced and integrated herd of sheep and goats (44.7%) and numerous pigs (33.2%) as recommended by the Roman writers on farming. Characteristic of this late period was a well populated chicken house (9.3%). The raising of domesticated chickens, it should be noted, became common in Italy after the Roman conquests in the East (though much earlier the Greeks were known to sacrifice them—"a cock to Asclepius"—and Romans raised their "sacred chickens". Worthy of comment is the disappearance of any trace of deer at San Biagio. In this we may see a reflection of the loss of the forests to pastureland for the numerous sheep and goats. From the treeless meadows and pastures the hunter came home with the occasional rabbit (0.3%) or hare (0.3%). The aquatic turtle had strayed (2.8%) not too far from the Basento River.

First figure indicates amount of bones and/or fragments
Second figure indicates percentage of total member recovered

	Pizzica		Incoronata		S. Biagio	
Bos taurus Cow	132	16.1	93	19.7	22	8.2
Bos t. brachyceros Cow (Sacrificial)	5	1.3	—		—	
Ovis vel Capra Sheep or Goat	43	11.7	170	36.0	120	44.7
Cervus elaphus Deer	33	9.0	36	7.6	—	
Sus scrofa Pig	2	0.5	—		—	
Sus domesticus Pig	38	10.4	133	28.2	89	33.2
Equus caballus Horse	60	14.4	10	2.1	2	0.7
Canis familiaris Dog	31	8.5	15	3.1	1	0.3
Canus vulpes Fox	2	0.5	—		—	
Lepus europaeus Rabbit	—		1	0.2	—	
Oryctolagus cuniculus/Rabbit			—		1	0.3

	Pizzica		Incoronata		S. Biagio	
Arvicola italicus Field Mouse	4	1.0	—		—	
Epimis rat. aless. Rat	—		—		1	0.3
Bufo bufo Toad	1	0.2	—		—	
Rana cfr. esculenta Frog	1	0.2	—		—	
Emys orbicularis Fresh Water Tortoise	—		12	2.5	5	2.8
Testudo cfr. hermanni/Tortoise	1	0.2	—		—	
Zamenius ssp. Black Snake	3	0.8	—		—	
Aegypius monachus Vulture	7	1.9	—		—	
Gallus domesticus Chicken	1	0.2	—		25	9.3
Corvus corax Raven	1	0.2	—		—	
Homo sapiens s.* Man	54				36	
Totals	365	100%	471	100%	268	100%

*Human remains are not considered in the percentages in the table.

1982

Barbara Bini, *Photographer, Rome, Italy*
Jennifer Brehob, *Ceramic Studies, UT at Austin*
Cesare D'Annibale, *Survey Leader, Brock University, Ontario*
Professor Robert L. Folk, *Geology Faculty, UT at Austin*
Sarah Green, *Cataloguer, Cambridge University, England*
Nancy Jircik, *Art History Faculty, University of St. Thomas, Houston*
Suzanne Jorgenson, *Artist, Sidney University, Australia*
Sandra Langston, *Artist, UT at Austin*
Alessandro Mineo, *University of Torino, Italy*
Ann Patterson, *Architect, UT at Austin*
Claes Petersson, *Survey, University of Stockholm*
Marianne Prohászka, *Metal Studies, University of Göteborg, Sweden*
Mark Reed, *Photographer, UT at Austin*
Pontus Reimers, *Survey, University of Stockholm*
Gene Storie, *Survey, UT at Austin*

EXCAVATION CREW

Leonardo Pacciani, *Specialist, Heavy Equipment, Bernalda*
Carmine Gallitelli, *Bernalda*
Rocco Paradiso, *Bernalda*
Umberto Sgrafetto, *Bernalda*
Francesca Quarato, *Restorer, Metaponto*

REPRESENTATIVE OF THE SUPERINTENDENCY

Giuseppe Di Taranto, *Montescaglioso*

CONSULTANTS

Ditta Chiartano, Metaponto
Dr. Lorenzo Costantini, *Paleobotany, Ministero dei Beni Culturali ed Ambientali, Rome*
Dr. Carolyn Koehler, *Amphorae, Classics Faculty, University of Maryland*
Don Sullivan, *Palynologist, University of California, Berkeley*

ADMINISTRATIVE SUPPORT, AUSTIN, 1981–1982

Ms. Leoda Anderson, *Classics, Administrative Assistant*
Professor G. K. Galinsky, *Classics, Chairman*
Dean Robert King, *Liberal Arts*
Dr. G. Fonken, *Vice-President for Research*

1981

Nanita Barchi, *Draftsman, University of Pennsylvania*
Deena Berg, *Architect, UT at Austin*
Jennifer Brehob, *Ceramic Studies, UT at Austin*
Dr. Lucilla Burn, *Ceramic Studies, Oxford University*
Boyce Cabaniss, *Faunal Studies, UT at Austin*
Cesare D'Annibale, *Survey Leader, Brock University, Ontario*
Pamela Dyson, *Survey, Washington, D.C.*
Professor Ingrid Edlund, *Classics Faculty, UT at Austin*
Professor Robert L. Folk, *Geology Faculty, UT at Austin*
Gloria Howell, *Architect, UT at Austin*
Anna Keys, *Ceramic Studies, UT at Austin*
Charles Pennington, *Trench Supervisor, Haverford College*
Ellen Simmons, *Draftsman, UT at Austin*
Chris Williams, *Photographer, UT at Austin*
Stephan Wolohojian, *Survey, Rutgers University*

EXCAVATION CREW

Alfredo Gallitelli, *Specialist, Excavation, Bernalda*
Leonardo Pacciani, *Specialist, Heavy Equipment, Bernalda*
Domenico D'Anzi, *Bernalda*
Carmine Gallitelli, *Bernalda*
Rocco Paradiso, *Bernalda*
Francesca Quarato, *Restorer, Metaponto*
Vittoria Quarato, *Restorer, Metaponto*

REPRESENTATIVE OF THE SUPERINTENDENCY

Giuseppe Di Taranto, *Montescaglioso*

CONSULTANTS

Ditta Chiartano, *Metaponto*
Dr. Lorenzo Costantini, *Paleobotany, Ministero dei Beni Culturali ed Ambientali, Rome*
Dr. John Hayes, *Ceramic Studies, Royal Ontario Museum*
Professor Tom Boyd, *Architecture, Art History Faculty, UT at Austin*